PROGRAMMABLE LOGIC CONTROLLERS

FRANK D. PETRUZELLA

GLENCOE

Macmillan/McGraw-Hill

Lake Forest, Illinois   Columbus, Ohio
Mission Hills, California   Peoria, Illinois

Sponsoring Editor: **Paul Sobel/Brian Mackin**
Editing Supervisor: **Carole Burke**
Design and Art Supervisor: **Meri Shardin/Nancy Axelrod Sharkey**
Production Supervisor: **Catherine Bokman**

Text Designer: **Levavi & Levavi**
Cover Designer: **Meri Shardin/Peri Zules**
Cover Photographer/Illustrator: **James Nazz**
Technical Studio: **York Graphic Services, Inc.**

Library of Congress Cataloging-in-Publication Data

Petruzella, Frank D.
    Programmable logic controllers/Frank D. Petruzella.
        p.   cm.
    Includes index.
    ISBN 0-07-049687-0
    1. Programmable controllers.   I. Title.
TJ223.P76P48 1989   88-13756
629.8'95—dc19   CIP

The manuscript and line art for this book were processed electronically.

PROGRAMMABLE LOGIC CONTROLLERS

Imprint 1992
Copyright © 1989 by the Glencoe Division of Macmillan/McGraw-Hill Publishing Company. All
rights reserved. Copyright © 1989 by McGraw-Hill, Inc. All rights reserved. Printed in the
United States of America. Except as permitted under the United States Copyright Act of 1976,
no part of this publication may be reproduced or distributed in any form or by any means, or
stored in a database or retrieval system, without the prior written permission of the publisher.
Send all inquiries to: Glencoe Division, Macmillan/McGraw-Hill, 936 Eastwind Drive, Westerville,
Ohio 43081.

4 5 6 7 8 9 10 11 12 13 14 15 A-HAL 00 99 98 97 96 95 94 93 92

ISBN 0-07-049687-0

CONTENTS

PREFACE

Programmable logic controllers (PLCs) are used in virtually every segment of industry where automation is required. They represent one of the fastest-growing segments of the industrial electronics industry. Since their inception, PLCs have proved to be the salvation of many manufacturing plants which previously relied on electromechanical control systems.

This text offers introductory students a first course in PLCs. It focuses on the underlying principles of how PLCs work and provides practical information about installing, programming, and maintaining a PLC system. No previous knowledge of PLC systems or programming is assumed. Employment opportunities in this field are very good, and this book is designed to help students acquire the necessary qualifications for these jobs.

In many instances the only source of information concerning programmable controllers is the user's manual published by the manufacturer. This textbook is not intended to replace the manufacturer's user's manual, but rather complement and expand on information found in the user's manual. The book discusses PLCs in a generic sense, and the content is broad enough to allow the information to be used with a wide range of PLC models. It is a highly adaptable text that can be used in many curriculums.

The text is written in an easy-to-read and understandable language with many clear illustrations to assist the student in comprehending the fundamentals of a PLC system. Objectives are listed at the beginning of each chapter to inform the student what will be learned. This list is followed by the subject material, which is generic in nature. The relay equivalent of the programmed instruction is explained first, followed by the appropriate PLC instruction. Each chapter concludes with a set of review questions and problems. The review questions are very closely related to the chapter objectives and should help students evaluate their understanding of the chapter. The problems range from easy to difficult, thus challenging students at various levels of competence. The answers to all chapter review questions and problems are found in the instructor's guide, which is available for use with this text.

All topics are covered in small segments, developing a firm foundation for each concept and operation before advancing to the next. An entire chapter is devoted to logic circuits as they apply to PLCs. PLC safety procedures and considerations are stressed throughout the text. Technical terms are defined when they are first used, and an extensive glossary provides easy referral to PLC terms. General troubleshooting procedures and techniques are stressed, and the student is instructed in how to analyze PLC problems in a systematic manner.

Both an activities manual and an instructor's guide are available for use with this text. The activities manual contains true/false, completion, matching, and multiple-choice test questions for each chapter of the text.

The best way to really understand any given PLC is to work with that PLC. Therefore, in the activities manual, each chapter contains a wide range of generic programming assignments and exercises for student practice with the PLC. The instructor's guide contains answers to all textbook review questions and problems as well as answers to the activities manual test questions.

I hope that you find the material presentation simple to read and understand, as well as informative.

Frank D. Petruzella

1

PROGRAMMABLE LOGIC CONTROLLERS (PLCs): AN OVERVIEW

Upon completion of this chapter you will be able to:

- Define what a programmable logic controller (PLC) is and list its advantages over relay systems
- Identify the main parts of a PLC and describe their functions
- Outline the basic sequence of operation for a PLC
- Idenfity the general classifications of PLCs according to the number of inputs and outputs and the size of the memory

1-1 PROGRAMMABLE LOGIC CONTROLLERS

A *programmable logic controller (PLC)* is a solid-state device designed to perform logic functions previously accomplished by electromechanical relays (Fig 1-1).

The design of most PLCs is similar to that of a computer. Basically, the PLC is an assembly of solid-state digital logic elements designed to make logical decisions and provide outputs. Programmable logic controllers are used for the control and operation of manufacturing process equipment and machinery.

Fig. 1-1 Programmable logic controller. *(Courtesy of Omron Electronics, Inc.)*

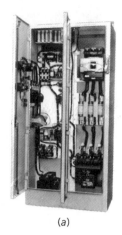

(a)

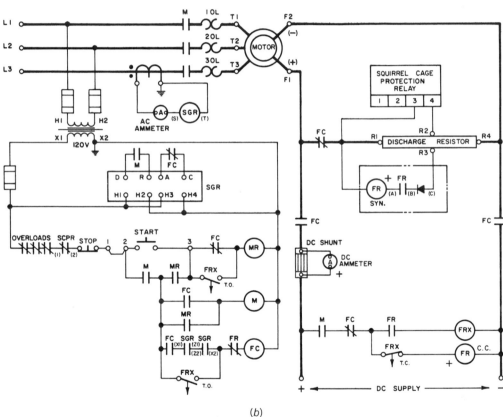

(b)

Fig. 1-2 (*a*) Typical hard-wired controller panel. (*b*) Typical hard-wired wiring diagram.   (*Courtesy of Allen-Bradley Company, Inc.*)

Programmable controllers offer several advantages over a conventional relay type of control. Relays have to be hard-wired to perform a specific function (Fig. 1-2). When the system requirements change, the relay wiring has to be changed or modified. In extreme cases, such as in the auto industry, complete control panels had to be replaced since it was not economically feasible to rewire the old panels with each model changeover. The programmable controller has eliminated much of the hand wiring associated with conventional relay control circuits. It is small and inexpensive compared to equivalent relay-based process control systems. Programmable controllers also offer solid-state reliability, lower power consumption, and ease of expandability.

1-2 PARTS OF A PLC

A typical PLC can be divided into three parts as illustrated in Fig. 1-3. These three components are the *central processing unit (CPU),* the *input/output (I/O) section,* and the *programming device.*

The CPU is the ''brain'' of the system (Fig. 1-4). Internally it contains various logic gate circuits. The CPU is a microprocessor-based system that replaces control relays, counters, timers, and sequencers. It is designed so that the user can enter the desired circuit in relay ladder logic. The CPU accepts (reads) input data from various sensing devices, executes the stored user program from memory, and sends appropriate output commands

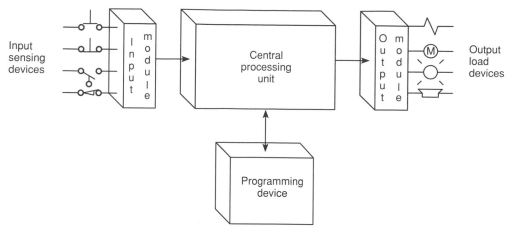

Fig. 1-3 PLC parts.

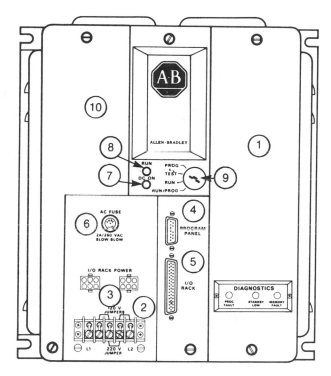

Legend:

1. Access memory and processor modules by removing panel

2. User power connections

3. I/O rack power socket

4. Program panel socket

5. I/O rack socket

6. Main input fuse

7. DC ON indicator

8. RUN indicator

9. Four-position mode select switch

10. System power supply module located here

Fig. 1-4 Typical processor unit.   *(Courtesy of Allen-Bradley Company, Inc.)*

to control devices. A direct current (dc) power source is required to produce the low-level voltage used by the processor and the I/O modules. This power supply can be housed in the CPU unit or may be a separately mounted unit, depending on the PLC system manufacturer.

The I/O section consists of input modules and output modules (Fig. 1-5). The I/O system forms the interface by which field devices are connected to the controller. The purpose of this interface is to condition the various signals received from or sent to external field devices. Input devices such as push buttons, limit switches, sensors, selector switches, and thumbwheel switches are hard-wired to terminals on the input modules. Output devices such as small motors, motor starters, solenoid valves, and indicator lights are hard-wired to the terminals on the output modules. These devices are also referred to as "field" or "real world" inputs and outputs. The terms *field* or *real world* are used to distinguish actual external devices that exist and must be physically wired from the internal user program that duplicates the function of relays, timers, and counters.

The programming device, or terminal, is used to enter the desired program into the memory of the processor. This program is entered using *relay ladder logic*. The program determines the sequence of operation and ultimate control of the equipment or machinery. The programming device must be connected to the controller only

when entering or monitoring the program. By designing the controller to be "electrician friendly," the PLC can be programmed by people without extensive computer programming experience. Actual programming is usually achieved by pushing keys on a keyboard. The programming device may be a hand-held unit with a light-emitting diode (LED) display (Fig.1-6*a*) or a desktop unit with a cathode-ray tube (CRT) display (Fig. 1-6*b*).

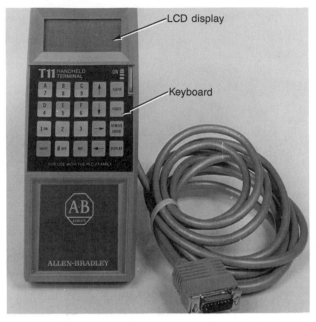

(a)

(b)

Fig. 1-6 Programming devices. *(a) (Courtesy of Allen-Bradely) (b) (Courtesy of Honeywell, Inc.)*

Fig. 1-5 Typical input and output.   *(Courtesy of Grayhill, Inc., La Grange, Illinois)*

1-3   PRINCIPLES OF OPERATION

To get an idea of how a PLC operates, consider the simple process control problem illustrated in Fig. 1-7. Here

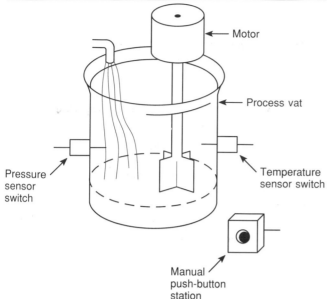

Fig. 1-7 Mixer process control problem.

a mixer motor is to be used to automatically stir the liquid in a vat when the temperature and pressure reach preset values. In addition, direct manual operation of the motor is provided by means of a separate push-button station. The process is monitored with temperature and pressure sensor switches that close their respective contacts when conditions reach their preset values.

This control problem can be solved using the relay method for motor control shown in the relay ladder diagram of Fig. 1-8. The motor starter coil (M) is energized when both the pressure and temperature switches are closed or when the manual push button is pressed.

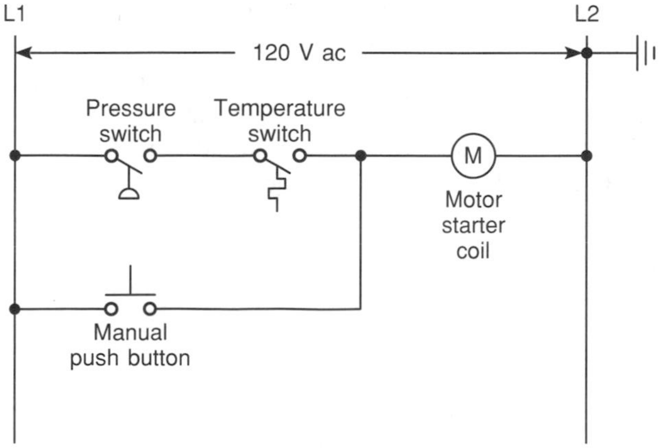

ical wiring connections for a 120-V ac input module are shown in Fig. 1-9.

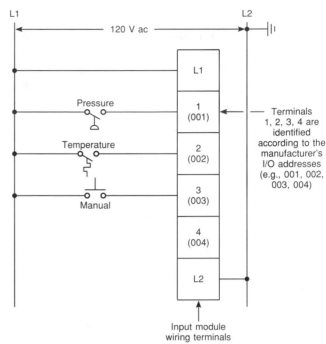

Fig. 1-9 Typical input module wiring connections.

The same output field device (motor starter coil) would also be used. This device would be hard-wired to an appropriate output module according to the manufacturer's labeling scheme. Typical wiring connections for a 120-V ac output module are shown in Fig. 1-10.

Fig. 1-8 Process control relay ladder diagram.

Now let's look at how a PLC might be used for this application. The same input field devices (pressure switch, temperature switch, and push button) are used. These devices would be hard-wired to an appropriate input module according to the manufacturer's labeling scheme. Typ-

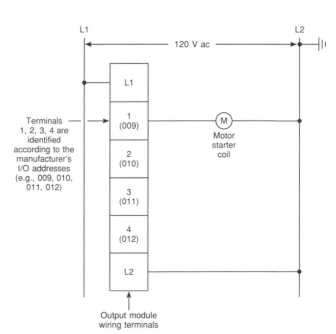

Fig. 1-10 Typical output module wiring connection.

Next, the PLC ladder logic diagram would be constructed and programmed into the memory of the CPU. A typical ladder logic diagram for this process is shown in Fig. 1-11. The format used is similar to the layout of the hard-wired relay ladder circuit. The individual symbols represent *instructions* while the numbers represent the instruction *addresses*. When programming the controller, these instructions are entered one by one into the processor memory from the operator terminal keyboard. Instructions are stored in the user program portion of the processor memory.

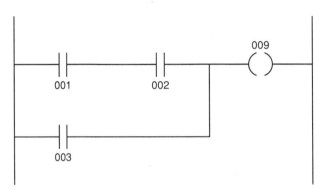

Note: Numbers 001, 002, 003, and 009 are identified with the pressure switch, temperature switch, manual push button, and motor starter coil, respectively.

Fig. 1-11 Process control PLC ladder logic diagram.

To operate the program, the controller is placed in the RUN mode, or operating cycle. During each operating cycle, the controller examines the status of input devices, executes the user program, and changes outputs accordingly. Each -| |- can be thought of as a set of normally open (NO) contacts. The -()- can be considered to represent a coil that, when energized, will close a set of contacts. In the ladder logic diagram of Fig. 1-11, the coil 009 is energized when contacts 001 and 002 are closed or when contact 003 is closed. Either of these conditions provides a continuous path from left to right across the rung that includes the coil.

The RUN operation of the controller can be described by the following sequence of events. First, the inputs are examined and their status is recorded in the controller's memory (a closed contact is recorded as a signal that is called a logic 1 and an open contact by a signal that is called a logic 0). Then the ladder diagram is evaluated, with each internal contact given OPEN or CLOSED status according to the record. If these contacts provide a current path from left to right in the diagram, the output coil memory location is given a logic 1 value and the output module interface contacts will close. If there is no conducting path on the program rung, the output coil memory location is set to logic 0 and the output module interface contacts will be open. The completion of one

cycle of this sequence by the controller is called a *scan*. The *scan time,* the time required for one full cycle, provides a measure of the speed of response of the PLC.

1-4 MODIFYING THE OPERATION

As mentioned, one of the important features of a PLC is the ease with which the program can be changed. For example, assume that our original process control circuit for the mixing operation must be modified as shown in the *relay* ladder diagram of Fig. 1-12. The change requires that the manual push-button control should be permitted to operate at any pressure *but not unless* the specified temperature setting has been reached.

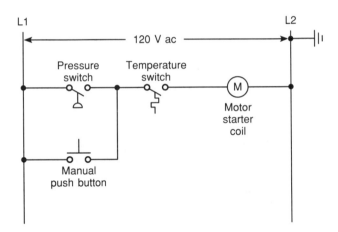

Fig. 1-12 Relay ladder diagram for modified process.

If a relay system were used, it would require some rewiring of the system (as shown in Fig. 1-12) to achieve the desired change. However, if a PLC system were used, *no* rewiring would be necessary. The inputs and outputs are still the same. All that is required is to change the PLC ladder logic diagram as shown in Fig. 1-13.

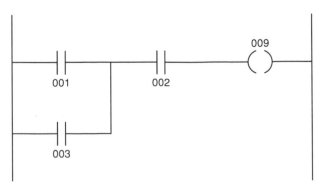

Fig. 1-13 PLC ladder logic diagram for modified process.

1-5 PLCs VERSUS COMPUTERS

The architecture of a PLC is basically the same as that of a general purpose computer. However, some important characteristics distinguish PLCs from general purpose computers. First, unlike computers, the PLC is designed to operate in the industrial environment with wide ranges of ambient temperature and humidity. A well-designed PLC is not usually affected by the electrical noise that is inherent in most industrial locations.

A second distinction between PLCs and computers is that the hardware and software of PLCs are designed for easy use by plant electricians and technicians. Unlike the computer, the PLC is programmed in *relay ladder logic* or other easily learned languages.

Computers are complex computing machines capable of executing several programs or tasks simultaneously and in any order. Most PLCs, on the other hand, execute a single program in an orderly and sequential fashion from first to last instruction.

Perhaps the most significant difference between a PLC and a computer is the fact that PLCs have been designed for installation and maintenance by plant electricians who are not required to be highly skilled computer technicians. Troubleshooting is simplified by the design of most PLCs in that they include fault indicators and written fault information that is displayed on the programmer screen. The modular interfaces for connecting the field devices are actually a part of the PLC and are easily connected and replaced.

1-6 PLC SIZE AND APPLICATION

There is much variation in size identification of PLCs. Typically PLCs are divided into three major size categories: small, medium, and large, each with distinct op-

Unlike computers, the programmable logic controller is designed to operate in the industrial environment with wide ranges of ambient temperature and humidity. *(Courtesy of Westinghouse Electric Corporation)*

Programmable logic controllers are available in a variety of sizes in units capable of providing simple to advanced levels of machine control. *(Courtesy of Westinghouse Electric Corporation)*

erating features. The small size category covers units with up to 128 I/Os and memories up to 2 K bytes. These PLCs are capable of providing simple to advanced levels of machine control.

Medium-size PLCs have up to 2048 I/Os and memories up to 32 K bytes. Special I/0 modules make medium PLCs adaptable to temperature, pressure, flow, weight, position, or any type of analog function commonly encountered in process control applications.

Large PLCs, of course, are the most sophisticated units of the PLC family. They have up to 8192 I/Os and memories up to 750 K bytes. PLCs of this size have virtually unlimited applications. Large PLCs can control individual production processes or entire plants.

The key factor in selecting a PLC is establishing exactly what the unit is supposed to do. In general it is not advisable to buy a PLC system that is larger than current needs dictate. However, future conditions should be anticipated to ensure that the system is the proper size to fill the current and possibly future requirements of an application.

Since its invention, the PLC has been successfully applied in virtually every segment of industry. This list includes steel mills, paper and pulp plants, chemical and automotive and power plants. Programmable logic controllers perform a great variety of control tasks, from repetitive ON/OFF control of a simple machine to sophisticated manufacturing and process control.

Gould's Micro 84 programmable controller includes the capability for analog and BCD register parameters, as well as providing communications with other programmable controllers and computers. *(Courtesy of Gould Industrial Automation Systems)*

REVIEW QUESTIONS

1. Define *programmable logic controller.*

2. List four advantages that PLCs offer over the conventional relay type of control system.

3. Describe the main function of each of the following main component parts of a PLC.
   (a) CPU
   (b) I/O modules
   (c) Programming device

4. Answer the following with reference to the unit process control relay ladder diagram of Fig. 1-8:
   (a) When do the pressure switch contacts close?
   (b) When do the temperature switch contacts close?
   (c) How are the pressure and temperature switches connected with respect to each other?
   (d) Describe the two conditions under which the motor starter coil will become energized.
   (e) What is the approximate value of the voltage drop across each of the following when their contacts are open?
      1. Pressure switch
      2. Temperature switch
      3. Manual push button

5. Answer the following with reference to the unit process control PLC ladder logic diagram of Fig. 1-11:
   (a) What do the individual symbols represent?
   (b) What do the numbers represent?
   (c) What is the number 002 identified with?
   (d) What is the number 009 identified with?
   (e) What two conditions will provide a continuous path from left to right across the rung?
   (f) Describe the sequence of operation of the controller for one scan of the program.

6. Compare the method by which the process control operation is changed in a relay system to the method for a PLC system.

7. Compare the PLC and general purpose computer with regard to:
   (a) Operating environment
   (b) Method of programming
   (c) Execution of program
   (d) Maintenance

8. (a) Explain the three size classifications for PLCs and state one general application for each size.
   (b) What are the two key factors in selecting the size of a PLC?

PROBLEMS

1. Given two single-pole switches, write a program that will turn on an output when both switch A and switch B are closed.
2. Given two single-pole switches, write a program that will turn on an output when either switch A or switch B is closed.
3. Given four NO push buttons (A-B-C-D), write a program that will turn a lamp on if push buttons A and B or C and D are closed.
4. Write a program for the relay ladder diagram shown in Fig. 1-14.
5. Write a program for the relay ladder diagram shown in Fig. 1-15.

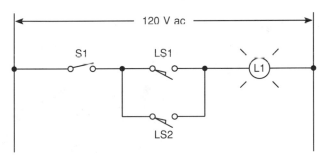

Fig. 1-14

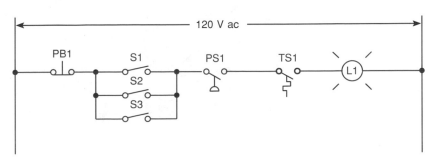

Fig. 1-15

2

PLC HARDWARE COMPONENTS

Upon completion of this chapter you will be able to:

■ List and describe the function of the hardware components used in PLC systems
■ Describe the basic circuitry and applications for discrete and analog I/O modules, and interpret typical I/O and CPU specifications

2-1 THE I/O SECTION

The input and output interface modules provide the equivalents of eyes, ears, and tongue to the brain of a PLC, the CPU. The I/O section consists of an I/O rack and individual I/O modules similar to that shown in Fig. 2-1. Input interface modules accept signals from the machine or process devices (120 V ac) and convert them into signals (5 V dc) that can be used by the controller. Output interface modules convert controller signals (5 V dc) into

Fig. 2-1 I/O rack and processor. *(Courtesy of Cincinnati Milacron)*

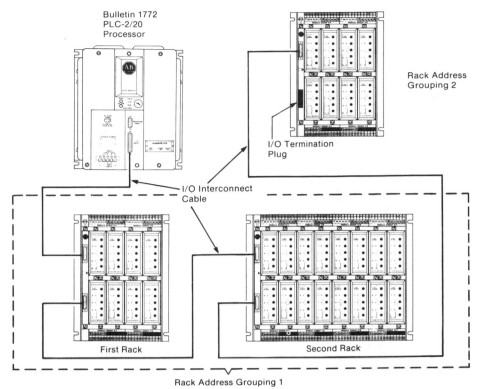

Fig. 2-2 Interconnecting cables between I/O racks and processor unit. *(Courtesy of Allen-Bradley Company, Inc.)*

external signals (120 V ac) used to control the machine or process.

In large PLC systems, I/O subsystems can be remotely located from the CPU. A remote subsystem is usually a rack-type enclosure in which the I/O modules are installed. An interconnecting cable allows communication between the processor and the remote I/O rack. This is shown in Fig. 2-2).

The location of a module within a rack and the terminal number of a module to which an input or output device is connected will determine the device's address (Fig. 2-3). Each input and output device must have a specific address. This address is used by the processor to identify where the device is located in order to monitor or control it. In addition, there is some means of connecting field wiring on the I/O module housing. Connecting the field wiring to the I/O housing allows easier disconnection and reconnection of the wiring in order to change modules. Lights are also added to each module to indicate the ON or OFF status of each I/O circuit. Most output modules also have blown fuse indicators.

A standard I/O module consists of a printed circuit board and a terminal assembly similar to that shown in Fig. 2-4. The printed circuit board contains the electronic circuitry used to interface the circuit of the processor with that of the input or output device. It is designed to plug into a slot or connector in the I/O rack or directly into the processor. The terminal assembly, which is attached to the front edge of the printed circuit board, is used for making field-wiring connections.

2-2 DISCRETE I/O MODULES

The most common type of I/O interface module is the discrete type. This type of interface connects field input devices of the ON/OFF nature such as selector switches, push buttons, and limit switches. Likewise, output control is limited to devices such as lights, small motors, solenoids, and motor starters, that require simple ON/OFF switching.

Each discrete I/O module is powered by some field-supplied voltage source. Since these voltages can be of different magnitude or type, I/O modules are available at various ac and dc voltage ratings as listed in Table 2-1.

Table 2-1 COMMON RATINGS FOR DISCRETE I/O INTERFACE MODULES

Input Interfaces	Output Interfaces
24 V ac/dc	12–48 V ac
48 V ac/dc	120 V ac
120 V ac/dc	230 V ac
230 V ac/dc	120 V dc
5 V dc (TTL level)	230 V dc
	5 V dc (TTL level)

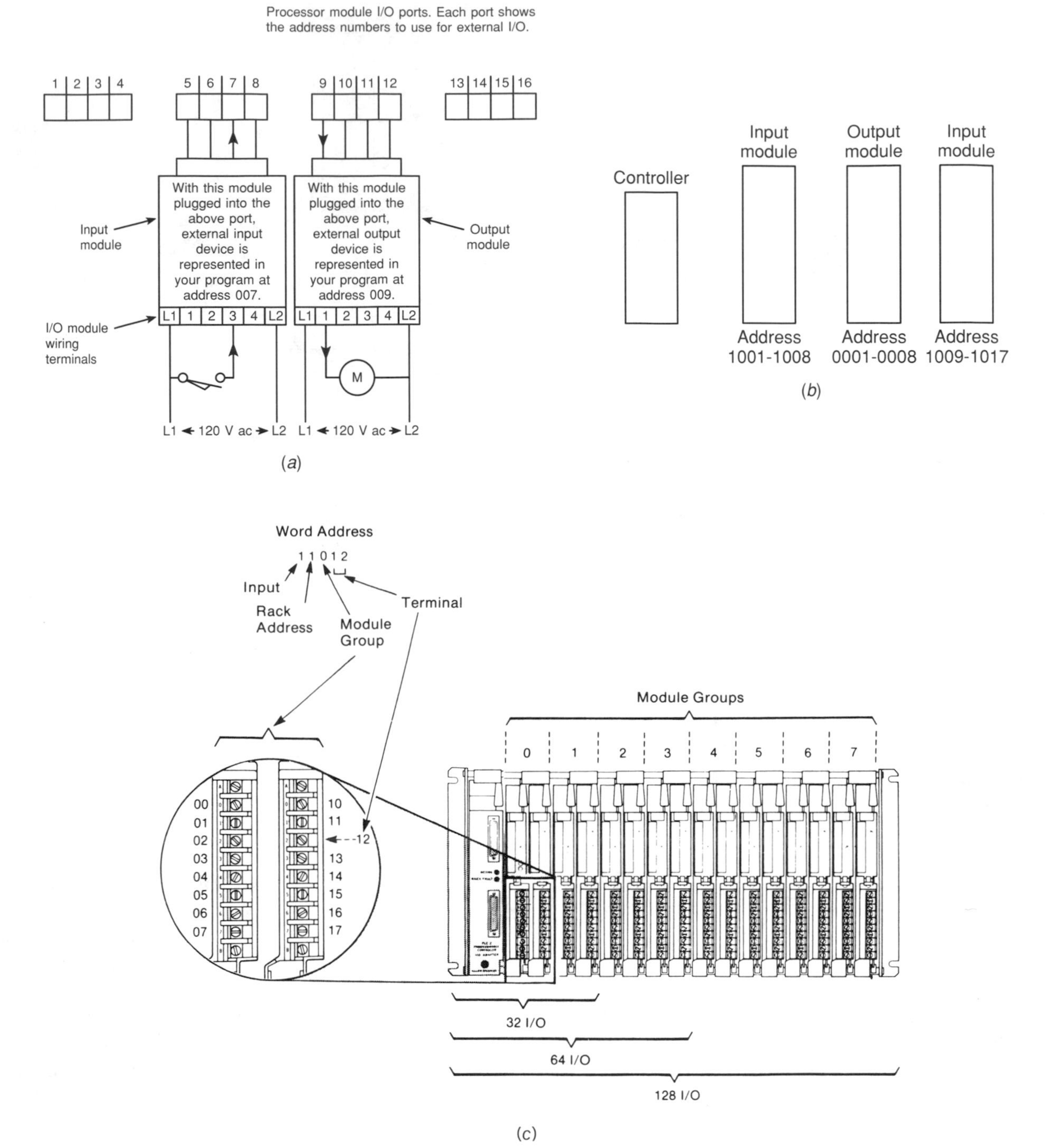

Processor module I/O ports. Each port shows
the address numbers to use for external I/O.

(a)

(b)

(c)

Fig. 2-3 I/O module address. (*a*) Three-digit. (*b*) Four-digit. (*c*) Five-digit.   (*Courtesy of Allen-Bradley Company, Inc.*)

Figure 2-5 shows a block diagram for one input of a typical alternating current (ac) interface input module. The input circuit is composed of two basic sections: the *power* section and the *logic* section. The power and logic sections are normally coupled together with a circuit, which electrically separates the two.

A simplified schematic and wiring diagram for one input of a typical ac input module is shown in Fig. 2-6*a* and *b*. When the push button is closed, 120 V ac is ap-

plied to the bridge rectifier through resistors R1 and R2. This produces a low-level direct current (dc) voltage, which is applied across the LED of the optical isolator. The zener diode (Z_D) voltage rating sets the minimum level of voltage that can be detected. When light from the LED strikes the phototransistor, it switches into conduction and the status of the push button is communicated in logic or low-level dc voltage to the processor. The optical isolator not only separates the higher ac in-

Fig. 2-4 I/O module construction. *(Courtesy of Reliance Electric Company)*

put voltage from the logic circuits, but also prevents damage to the processor due to line voltage transients. Optical isolation also helps reduce the effects of electrical noise, common in the industrial environment, which can cause erratic operation of the processor. Coupling and isolation can also be accomplished by use of a pulse transformer.

Figure 2-7 shows a block diagram for one output of a typical interface output module. Like the input module, it is composed of two basic sections: the *power* section and the *logic* section, coupled by an *isolation* circuit. The output interface can be thought of as a simple electronic switch to which power is applied to control the output device.

A simplified schematic and wiring diagram for one output of a typical ac output module is shown in Fig. 2-8a. As part of its normal operation, the processor sets the output status according to the logic program. When the processor calls for an output, a voltage is applied across the LED of the isolator. The LED then emits light, which switches the phototransistor into conduction. This in turn switches the *triode ac semiconductor switch (triac)* into conduction, which, in turn, turns on the lamp. Since the triac conducts in either direction, the output to the lamp is alternating current. The triac, rather than having ON and OFF status, actually has LOW and HIGH resistance levels, respectively. In its OFF state (HIGH

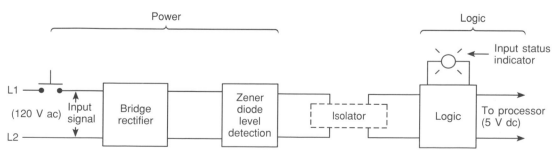

Fig. 2-5 Block diagram of ac interface input module.

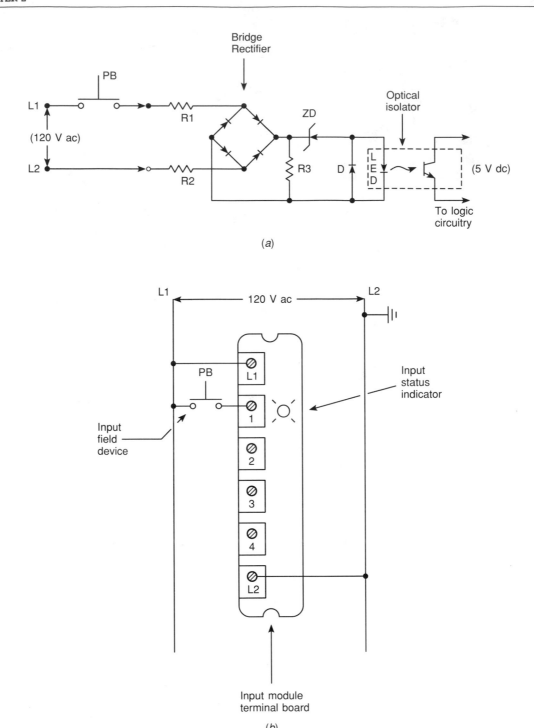

(a)

(b)

Fig. 2-6 (a) Simplified schematic for an ac input module. (b) Typical input module wiring connection.

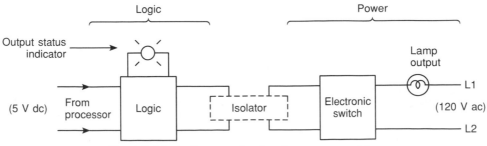

Fig. 2-7 Block diagram of ac interface output module.

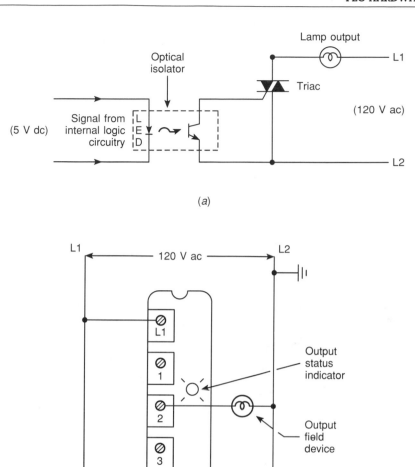

Fig. 2-8 (*a*) Simplified schematic for an ac output module. (*b*) Typical output module wiring connection.

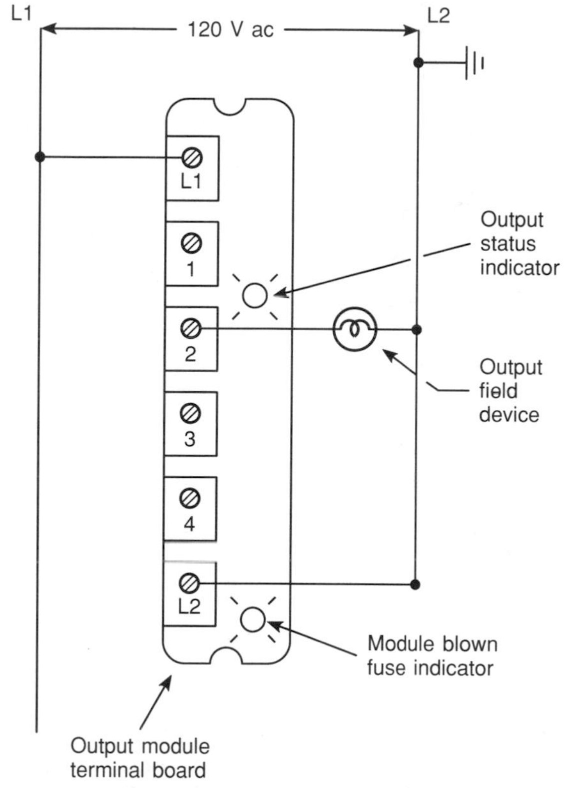

resistance), a small leakage current of a few milliamperes still flows through the triac. As with input circuits, the output interface is usually provided with LEDs that indicate the status of each output. In addition, if the module contains a fuse, a fuse status indicator may also be used (Fig. 2-8*b*).

Individual ac outputs are usually limited by the size of the triac to 2 or 3 amperes (A). The maximum current load for any one module is also specified. To protect the output module circuits, specified current ratings should not be exceeded. For controlling larger loads, such as large motors, a standard control relay is connected to the output module. The contacts of the relay can then be used to control a larger load or motor starter as shown in

Fig. 2-9. When a control relay is used in this manner, it is called an *interposing* relay.

2-3 ANALOG I/O MODULES

The earlier PLCs were limited to discrete I/O interfaces, which allowed only ON/OFF-type devices to be connected. This limitation meant that the PLC could have only partial control of many process applications. Today, however, a complete range of both discrete and analog interfaces are available that will allow controllers to be applied to practically any type of control process.

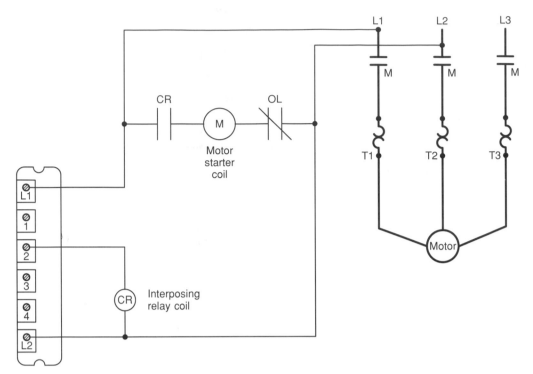

Fig. 2-9 Interposing relay connection.

Analog input interface modules contain the circuitry necessary to accept analog voltage or current signals from analog field devices. These inputs are converted from an analog to a digital value by an *analog-to-digital (A/D)* converter circuit. The conversion value, which is proportional to the analog signal, is expressed as a 12-bit binary or as a three-digit binary-coded decimal (BCD) for use by the processor. Analog input sensing devices

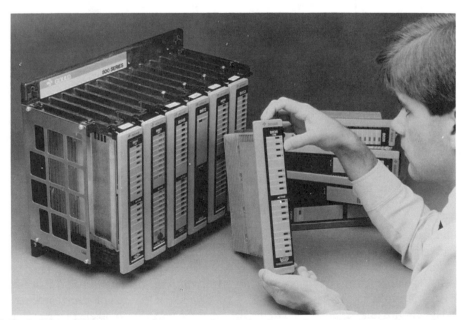

Today's I/O modules cover a wide range of control voltages, increasing the number and type of field devices that can be interfaced with them.   *(Courtesy of Gould Industrial Automation Systems)*

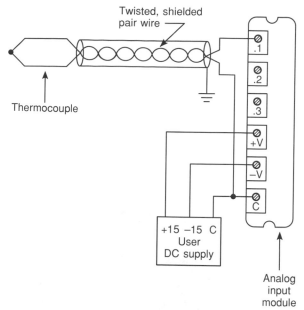

Fig. 2-10 Typical thermocouple connection to an analog input module.

include temperature, light, speed, pressure, and position transducers. Figure 2-10 shows a typical analog input interface module connection to a thermocouple. A varying dc voltage in the millivolt range, proportional to the temperature being monitored, is produced by the thermocouple. This voltage is amplified and digitized by the analog input module and then sent to the processor on command from a program instruction. Because of the low voltage level of the input signal, a shielded cable is used in wiring the circuit to reduce unwanted electrical noise signals that can be induced in the conductors from other wiring. This noise can cause temporary operating errors that can lead to hazardous or unexpected machine operation.

The analog output interface module receives from the processor digital data, which is converted into a proportional voltage or current to control an analog field device. The digital data is passed through a *digital-to-analog (D/A)* converter circuit to produce the necessary analog form. Analog output devices include small motors, valves, analog meters, and seven-segment displays.

2-4   I/O SPECIFICATIONS

Manufacturer's specifications provide much information about how an interface device is correctly and safely used. The specifications place certain limitations, not only on the module, but also on the field equipment that it can operate. The following is a list of some typical manufacturers I/O specifications along with a short description of what is specified.

Nominal input voltage   This ac or dc value specifies the magnitude and type of voltage signal that will be accepted

On-state input voltage range   This value specifies the voltage at which the input signal is recognized as being absolutely on.

Nominal current per input   This value specifies the minimum input current that the input devices must be capable of driving to operate the input circuit.

Ambient temperature rating   This value specifies what the maximum temperature of the air surrounding the I/O modules should be for best operating conditions.

Input delay   This value specifies the time duration for which the input signal must be on before being recognized as a valid input. This delay is a result of filtering circuitry provided to protect against contact bounce and voltage transients. This input delay is typically in the 9-ms to 25-ms range.

Nominal output voltage   This ac or dc value specifies the magnitude and type of voltage source that can be controlled by the output.

Output voltage range   This value specifies the minimum and maximum output operating voltages. An output circuit rated at 120 V ac, for example, may have an absolute working range of 92 V ac (min.) to 138 V ac (max.).

Maximum output current rating per output and module   These values specify the maximum current that a single output and the module as a whole can safely carry under load (at rated voltage).

Maximum surge current per output   This value specifies the maximum inrush current and duration (eg., 20 A for 0.1 s) for which an output circuit can exceed its maximum continuous current rating.

Off-state leakage current per output   This value specifies the maximum value of leakage current that flows through the output in its OFF state.

Electrical isolation   This maximum value (volts) defines the isolation between the I/O circuit and the controller's logic circuitry. Although this isolation protects the logic side of the module from excessive input or output voltages or current, the power circuitry of the module may be damaged.

2-5   THE CPU

The CPU houses the processor-memory module(s), communications circuitry, and power supply. Figure 2-11a is a simplified illustration of the CPU. Central processing unit architectures may differ from one manufacturer to another, but in general most of them follow this organization. The power supply may be located inside the CPU enclosure or may be a separate unit mounted next to the enclosure as shown in Fig. 2-11b. Depending upon the type of memory, volatile or nonvolatile, the power supply could also include a backup battery system.

The term *CPU* is often used interchangeably with the term *processor*. However, by strict definition, the *CPU* term encompasses all the necessary elements that form

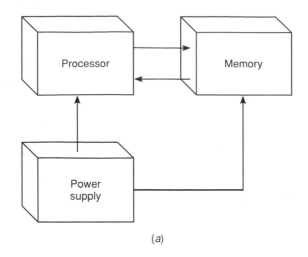

(a)

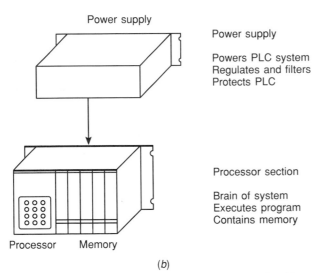

(b)

Fig. 2-11 Major components of the CPU. (a) Simplified illustration of the CPU. (b) Power supply mounted outside CPU enclosure.

the intelligence of the system. There are definite relationships between the sections that form the CPU and constant interaction among them. The processor is continually interacting with the system memory to interpret and execute the user program that controls the machine or process. The system power supply provides all the necessary voltage levels to ensure proper operation of all processor and memory components.

2-6 THE PROCESSOR-MEMORY MODULE

The processor-memory module that forms the major part of the CPU housing is the *brain* of the programmable controller. This module contains the microprocessor, memory chips, circuits that store and retrieve information from the memory, and communications circuits re-

quired for the processor to interface with the programming device. For smaller systems the microprocessor-memory and communications are all in one module similar to that shown in Fig. 2-12.

Fig. 2-12 Processor module. *(Courtesy of Square D Company)*

In recent years the decision-making capabilities of the PLC have gone far beyond simple logic processing. The processor may perform other functions such as timing, counting, latching, comparing, and complicated math beyond the basic four functions of addition, subtraction, multiplication, and division. These additional processor functions, found on larger PLCs, often necessitate the use of individual modules for such functions as memory, communication, and arithmetic.

2-7 MEMORY DESIGN

Memory is where the control plan or program is held or stored in the controller. The information stored in the memory relates to how the input and output data should be processed.

The complexity of the program determines the amount of memory required. Memory elements store individual pieces of information called *bits* (for *binary digits*). The amount of memory capacity is specified in increments of 1000 or in "K" increments, where 1 K is 1024 *bytes* of memory storage (a byte is 8 bits).

Although there are many types, memory can be placed into two categories: *volatile* and *nonvolatile*. Volatile memory will lose its stored information if all operating power is lost or removed. Volatile memory is easily altered and quite suitable for most applications when supported by battery backup.

Nonvolatile memory has the ability to retain stored information when power is removed accidentally or intentionally. Although nonvolatile memory generally is unalterable, there are special types used in which the stored information can be changed.

2-8 MEMORY TYPES

Today's PLCs make use of many different types of volatile and nonvolatile memory devices. Following is a generalized description of a few of the more common types. Details for specific memory types can be obtained from the specification sheets provided as part of the software package for a controller.

Read-Only Memory (ROM)

Read-only memory (ROM) is designed so that information stored in memory can only be read, and under ordinary circumstances cannot be changed. Information found in the ROM is placed there by the manufacturer, for the internal use and operation of the PLC. Read-only

memories are nonvolatile; they retain their information when power is lost and do not require battery backup.

Random Access Memory (RAM or R/W)

Random access memory (RAM or R/W), often referred to as *read-write (R/W) memory,* is designed so that information can be written into or read from the memory. There are two types of RAM: the volatile RAM and the nonvolatile RAM (see "Magnetic-Core Memory" on page 20). Today's controllers, for the most part, use the CMOS-RAM with battery support for user program memory. Random access memory provides an excellent means for easily creating and altering a program. The CMOS-RAM is becoming very popular because it has a very low current drain (15-μA range) when not being accessed, and the information stored in its memory can be retained by as little as 2 V dc.

Programmable Read-Only Memory (PROM)

The *programmable read-only memory (PROM)* is a special type of ROM. Programmable read-only memory allows initial and/or additional information to be written into the chip. Programmable read-only memory may be written into only once after being received from the manufacturer. Programming is accomplished by pulses of current that melt fusible links in the chip, preventing it from

RAM memory card. *(Courtesy of Klockner-Moeller Ltd.)*

being reprogrammed. Very few controllers use PROM for program memory since any program change would require a new set of PROM chips.

Erasable Programmable Read-Only Memory (EPROM)

The *erasable programmable read-only memory (EPROM)* is a specially designed PROM that can be reprogrammed after being entirely erased with the use of an ultraviolet light source. The EPROM chip has a quartz window over a silicon material that contains the integrated circuits (Fig. 2-13). This window is normally cov-

(a)

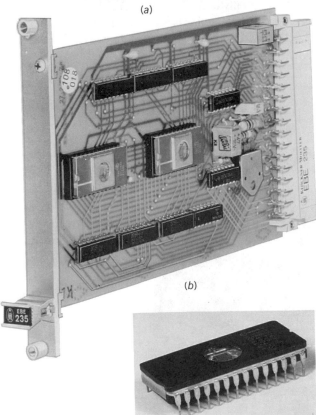

(b)

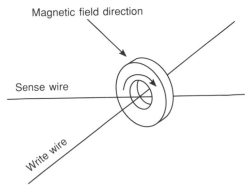

(c)

Fig. 2-13 Typical EPROM chip. *(a)* and *(b) Klockner-Moeller Ltd.) (c) (Allen-Bradley Company, Inc.)*

ered by an opaque material. When the opaque material is removed and the circuitry exposed to ultraviolet light for approximately 20 min, the memory content can be erased. Once erased, the EPROM chip can be reprogrammed using the programming device.

Electrically Alterable Read-Only Memory (EAROM)

Electrically alterable read-only memories (EAROMs) are similar to EPROMs, but do not require an ultraviolet light source to erase them. Instead, an erasing voltage is applied to the proper pin connections of the EAROM chip to wipe it clean. Once erased, the chip can be reprogrammed.

Electrically Erasable Programmable Read-Only Memory (EEPROM)

Electrically erasable programmable read-only memory (EEPROM) is a nonvolatile memory that offers the same programming flexibility as does RAM. Several small and medium-size controllers use EEPROM as the only memory for the system. It provides permanent storage of the program and can be easily changed using standard programming devices.

Magnetic-Core Memory

Magnetic-core memory uses small, doughnut-shaped ferrite cores for storing information (Fig. 2-14). Wires that carry electrical current are passed through the hole and are used to induce a magnetic field in the core. The direction of the current flow determines the direction of the induced magnetic field. The direction of the magnetic field can first be set (WRITE) and later be read (SENSE) to determine its logic status. Core memory is nonvolatile because power is not necessary to keep the

Magnetic field direction

Sense wire

Write wire

Fig. 2-14 Magnetic core memory.

cores magnetized. Core memory has the disadvantage of being bulky, slow, and relatively expensive. Core memory was used in many of the first PLCs and is still used in a few controllers.

2-9 PROGRAMMING DEVICES

Easy-to-use programming equipment is one of the important features of programmable controllers. The programming device provides the primary means by which the user can communicate with the circuits of the controller (Fig. 2-15).

The programming device allows the user to enter, change, or monitor a PLC controller program. Easy-to-use programming equipment is an important feature of the PLC. *Industrial CRT terminals* are perhaps the most commonly used devices for programming the controller. These terminals are self-contained video display units with a keyboard and the electronics necessary to communicate with the CPU and to display data. The CRT offers the advantage of displaying large amounts of logic on the screen, which simplifies the interpretation of the program (Fig. 2-16).

Miniprogrammers, also known as hand-held programmers, are an inexpensive and portable means for programming small PLCs. The display is usually an LED or liquid-crystal display (LCD) type, and the keyboard consists of numeric keys, programming instruction keys, and special function keys (Fig. 2-17).

2-10 PROGRAM LOADERS

Program loaders are used to record and store the user program or to load preprogrammed instructions into the processor. There are two types of program loaders: magnetic cassette or disk recorders and electronic memory modules. Cassette or disk recorders use magnetic tapes or disks to record and store the user program. Recording the user program on tape or disk provides a backup program in the event the processor program is lost as a result of memory failure or accidental erasure.

Electronic memory modules are the smallest storing and reloading devices currently available. These modules contain an EEPROM memory along with the electronics necessary to write or read a complete program into and from the module (Fig. 2-18).

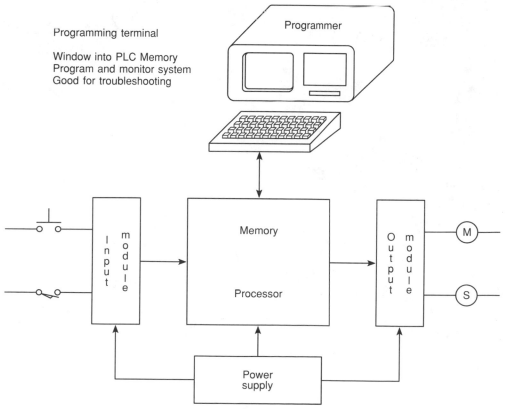

Fig. 2-15 User communications with PLC circuits.

Programming terminal provides window into PLC memory. *(Courtesy of Square D Company)*

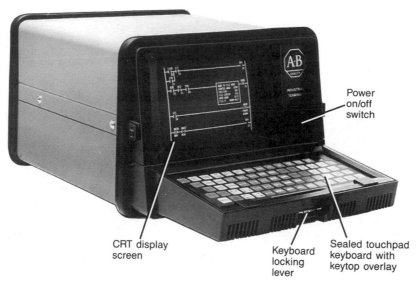

Fig. 2-16 A CRT industrial terminal. *(Courtesy of Allen-Bradley Company, Inc.)*

(a)

(b)

Fig. 2-17 Miniprogrammer. *(a) (Courtesy of Reliance Electric Company) (b) (Courtesy of EATON Corporation, Cutler-Hammer Products)*

1. Write-on area.

2. LED status indicator.

1. Write-on area: A write-on area is provided on the memory module, for use in identifying its contents.

2. LED status indicator: The LED status indicator on the front of the memory module will be ON when a program is being loaded into the module.

The EEPROM module installed.

Fig. 2-18 Electronic memory module. *(Courtesy of Allen-Bradley Company, Inc.)*

REVIEW QUESTIONS

1. What is the function of a PLC input module?

2. What is the function of a PLC output module?

3. How does the processor identify the location of a specific input or output device?

4. What type of field input devices are suitable for use with discrete input modules?

5. What type of field output devices are suitable for use with discrete output modules?

6. List three functions of the optical isolator circuit used in I/O module circuits.

7. Name the two basic sections of an I/O module.

8. What electronic component is often used as the switching device for 120-V ac output interface modules?

9. (a) What is the maximum current rating for a typical 120-V ac output interface module?
   (b) Explain how outputs with larger current requirements are handled.

10. (a) Compare discrete and analog I/O modules with respect to the type of input or output devices they can be used with.

(b) Explain the function of the A/D converter circuit used in analog input modules.
(c) List three common types of analog input sensing devices.
(d) Why is shielded cable often used when wiring low-voltage analog sensing devices?

11. Write a short description for each of the following I/O specifications:
   (a) Nominal input voltage
   (b) On-state input voltage range
   (c) Nominal current per input
   (d) Nominal output voltage
   (e) Output voltage range
   (f) Maximum output current rating
   (g) OFF-state leakage current per output
   (h) Electrical isolation

12. Explain the basic function of each of the three major parts of the CPU.

13. State three other functions, in addition to simple logic processing, that PLC processors are capable of performing.

14. Compare the memory storage characteristics of volatile and nonvolatile memory elements.

15. Compare ROM and RAM memory design with regard to:
    (a) How information is placed into the memory
    (b) How information in the memory is changed
    (c) Classification as volatile or nonvolatile

16. (a) How is initial and/or additional information written into a PROM chip?
    (b) What is the main limitation of PROM memory chips?

17. How is the program erased in the following chips?

    (a) EPROM   (b) EAROM   (c) EEPROM

18. (a) What determines the logic status of a single magnetic-core memory element?
    (b) How is the logic status changed?
    (c) What are two specific disadvantages of memory?

19. List three possible functions of a PLC programming device.

20. (a) What is the purpose of a PLC program loader?
    (b) What two types are available for use with PLCs?

PROBLEMS

1. A discrete 120-V ac output module is to be used to control a 230-V dc solenoid valve. Draw a diagram showing how this could be accomplished using an interposing relay.

2. Assume a thermocouple generates a linear voltage of from 20 mV to 50 mV when the temperature changes from 750°F to 1250°F. How much voltage will be generated when the temperature of the thermocouple is at 1000°F?

3. (a) The input delay time of a given module is specified as 12 ms. How much is this expressed in seconds?

    (b) The output leakage current of a given module is specified as 950 μA. How much is this expressed in amperes?
    (c) The maximum ambient temperature for a given I/O module is specified as 60°C. How much is this expressed in degrees Fahrenheit?

4. Create a typical five-digit address (according to Fig. 2-3) for each of the following?
    (a) A push button connected to terminal 5 of module group 2 located on rack 1.
    (b) A lamp connected to terminal 3 of module group 0 located on rack 2.

3

NUMBER SYSTEMS AND CODES

Upon completion of this chapter you will be able to:

- Define the decimal, binary, octal, and hexadecimal numbering systems and be able to convert from one numbering or coding system to another
- Explain the BCD and Gray code systems
- Define the terms *bit, byte, word, least significant bit (LSB)* and *most significant bit (MSB)* as they apply to binary memory locations
- Describe the purpose of the encoder and the decoder integrated-circuits (IC)

3-1 DECIMAL SYSTEM

Knowledge of different number systems and digital codes is quite useful when working with PLCs or with most any type of digital computer. This is true because a basic requirement of these devices is to represent, store, and operate on numbers. In general, PLCs work on binary numbers in one form or another; these are used to represent various codes or quantities.

The *decimal system,* which is most common to us, has a base of 10. The base of a number system determines the total number of different symbols or digits used by that system. For instance, in the decimal system, 10 unique numbers or digits—i.e., the digits 0 through 9— are used: the total number of symbols is the same as the base, and the symbol with the largest value is 1 less than the base.

The value of a decimal number depends on the digits that make up the number and the place value of each digit. A place (weight) value is assigned to each position that a digit would hold from right to left. In the decimal system the first position, starting from the rightmost position, is 0; the second is 1; the third is 2; and so on up to the last position. The weighted value of each position can be expressed as the base (10 in this case) raised to the power of the position. For the decimal system then, the position weights are 1, 10, 100, 1000, etc. Figure 3-1 illustrates how the value of a decimal number can be calculated by multiplying each digit by the weight of its position and summing the results.

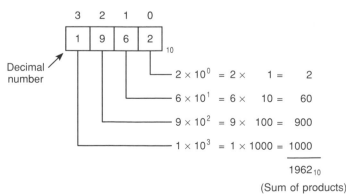

Fig. 3-1 Weighted value in the decimal system.

26

3-2 BINARY SYSTEM

The *binary system* uses the number 2 as the base. The only allowable digits are 0 and 1. With digital circuits it is easy to distinguish between two voltage levels (i.e., +5 V and 0 V), which can be related to the binary digits 1 and 0 (Fig. 3-2). Therefore the binary system is a very applicable one for PLCs and computer systems.

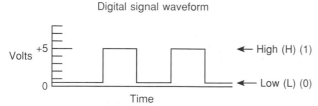

Fig. 3-2 Digital signal waveform.

Table 3-1 COUNTING TO DECIMAL 15 IN THE BINARY SYSTEM

Count	Decimal Number	Binary Number
Zero	0	00000
One	1	00001
Two	2	00010
Three	3	00011
Four	4	00100
Five	5	00101
Six	6	00110
Seven	7	00111
Eight	8	01000
Nine	9	01001
Ten	10	01010
Eleven	11	01011
Twelve	12	01100
Thirteen	13	01101
Fourteen	14	01110
Fifteen	15	01111

Since the binary system uses only two digits, each position of a binary number can go through only two changes, and then a 1 is carried to the immediate left position. Table 3-1 shows how binary numbers are used to count up to a decimal value of 15.

The decimal equivalent of a binary number is calculated in a manner similar to that used for a decimal number. This time the weighted values of the positions are 1, 2, 4, 8, 16, 32, 64 etc. Instead of being 10 raised to the power of the position, the weighted value is 2 raised to the power of the position. Figure 3-3 illustrates how the binary number 10101101 is converted to its decimal equivalent: 173.

Each digit of a binary number is known as a *bit*. In a PLC the processor-memory element consists of hundreds or thousands of locations. These locations, or *registers*, are referred to as *words*. Each word is capable of storing data in the form of *binary digits*, or *bits*. The number of bits that a word can store depends on the type of PLC system used. Eight-bit and sixteen-bit words are the most common. Bits can also be grouped within a word into *bytes*. Usually a group of 8 bits is a byte, and a group of 1 or more bytes is a word. Figure 3-4 illustrates a 16-bit word made up of 2 bytes. The *least significant bit (LSB)* is the digit that represents the smallest value and the *most significant bit (MSB)* is the digit that represents the largest value.

If a memory size is 884 words, then it can actually store 14,144 (884 x 16) bits of information using 16-bit words or 7072 (884 x 8) using an 8-bit word. Therefore,

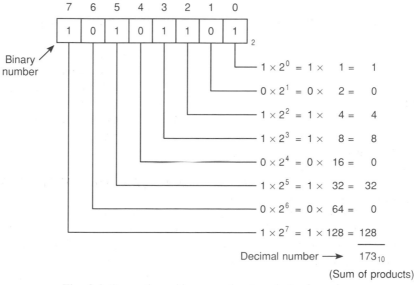

Fig. 3-3 Converting a binary number to a decimal number.

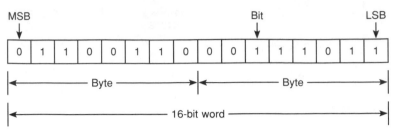

Fig. 3-4 A 16 bit word.

when comparing different PLC systems, one must know the number of bits per word of memory in order to determine the relative capacity of the systems' memories.

To convert a decimal number to its binary equivalent, we must perform a series of divisions by 2. Figure 3-5 illustrates the conversion of the decimal number 47 to binary. We start by dividing the decimal number by 2. If there is a remainder, it is placed in the LSB of the binary number. If there is no remainder, a 0 is placed in the LSB. The result of the division is brought down, and the process repeated until the result of successive divisions has been reduced to 0.

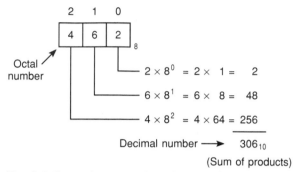

Fig. 3-6 Converting an octal number to a decimal number.

Table 3-2 BINARY AND RELATED OCTAL CODE

Binary	Octal
000	0
001	1
010	2
011	3
100	4
101	5
110	6
111	7

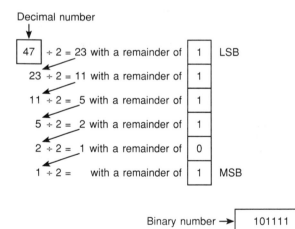

Fig. 3-5 Converting a decimal number to a binary number.

As mentioned, octal is used as a convenient means of handling large binary numbers. As shown in Table 3-2, one octal digit can be used to express three binary digits.

Therefore, the octal number 462 can be converted to its binary equivalent by assembling the 3-bit groups as illustrated in Fig. 3-7.

Thus, octal 462 is binary 100110010 and decimal 306. Notice the simplicity of the notation. The octal 462 is much easier to read and write than its binary equivalent. Most PLCs use the octal numbering system for referencing I/O and memory addresses.

3-3 OCTAL SYSTEM

To express the number in the binary system requires many more digits than in the decimal system. Too many binary digits can become cumbersome to read or write. To solve this problem the *octal numbering system* is brought into use. This system uses the number 8 as its base and makes use of eight digits: 0 through 7. As in all other number systems, each digit in an octal number has a weighted decimal value according to its position. Figure 3-6 illustrates how the octal number 462 is converted to its decimal equivalent: 306.

3-4 HEXADECIMAL SYSTEM

The *hexadecimal (hex) numbering system* provides even shorter notation than octal. Hexadecimal uses a base of 16. It employs 16 digits: numbers 0 through 9, and letters A through F, with A through F being substituted for numbers 10 to 15, respectively (Table 3-3).

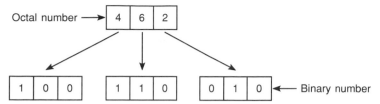

Fig. 3-7 Converting an octal number to a binary number.

Table 3-3   HEXADECIMAL NUMBERING SYSTEM

Hexadecimal	Binary	Decimal
0	0000	0
1	0001	1
2	0010	2
3	0011	3
4	0100	4
5	0101	5
6	0110	6
7	0111	7
8	1000	8
9	1001	9
A	1010	10
B	1011	11
C	1100	12
D	1101	13
E	1110	14
F	1111	15

Hexadecimal numbers can be expressed as their decimal equivalents by using the "sum of the weights" method. The place weights are powers of 16; Fig. 3-8 illustrates how the conversion would be done for the hex number 1B7.

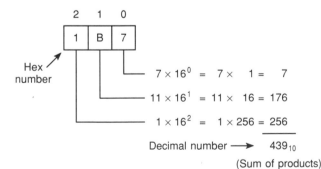

Fig. 3-8 Converting a hexadecimal number to a decimal number.

Like octal numbers, hexadecimal numbers can easily be converted to binary numbers. Conversion is accomplished by writing the 4-bit binary equivalent of the hex digit for each position, as illustrated in Fig. 3-9.

As Figs. 3-8 and 3-9 show, the hex number 1B7 is 00011011011 in binary and 439 in decimal.

The hexadecimal number system is used by some PLCs for entering output instructions into a sequencer.

3-5   BCD SYSTEM

The BCD system provides a convenient means to handle large numbers that need to be input to or output from a PLC. The BCD system represents decimal numbers as patterns of 1's and 0's. This system provides a means of converting a code readily handled by humans (decimal) to a code readily handled by the equipment (binary).

The BCD system uses 4 bits to represent each decimal digit. The 4 bits used are the binary equivalents of the numbers from 0 to 9. In the BCD system, the largest decimal number that can be displayed by any four digits is 9. Table 3-4 shows the 4-bit binary equivalents for each decimal number 0 through 9.

Table 3-4   FOUR-BIT BINARY EQUIVALENTS FOR DECIMAL NUMBERS 0 THROUGH 9

Decimal	BCD
0	0000
1	0001
2	0010
3	0011
4	0100
5	0101
6	0110
7	0111
8	1000
9	1001

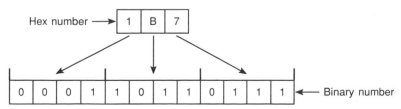

Fig. 3-9 Converting a hexadecimal number to a binary number.

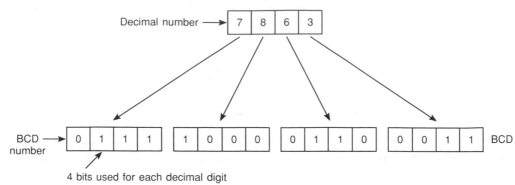

4 bits used for each decimal digit

Fig. 3-10 The BCD representation of a decimal number.

The BCD representation of a decimal number is obtained by replacing each decimal digit by its BCD equivalent. To distinguish the BCD numbering system from a binary system, a BCD designation is placed to the right of the units digit. The BCD representation of the decimal 7863 is shown in Fig. 3-10.

3-6 GRAY CODE

The Gray code has the advantage that for each "count" (each transition from one number to the next) *only one* digit changes. Table 3-5 shows the Gray code and the binary equivalent for comparison. In binary, as many as four digits could change for a single "count." For example, the transition from binary 0111 to 1000 (decimal 7 to 8) involves a change in all four digits. This kind of change increases the possibility for error in certain digital circuits.

Table 3-5   GRAY CODE AND BINARY EQUIVALENT

Gray Code	Binary
0000	0000
0001	0001
0011	0010
0010	0011
0110	0100
0111	0101
0101	0110
0100	0111
1100	1000
1101	1001
1111	1010
1110	1011
1010	1100
1011	1101
1001	1110
1000	1111

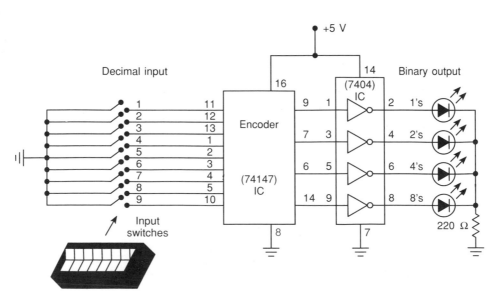

Fig. 3-11 Decimal to binary encoder.

3-7 ENCODING AND DECODING

The PLC, for the most part, uses digital integrated-circuit (IC) chips for conversion from one number system or code to another. An *encoder* IC is used to convert from *decimal* to *binary* numbers. Figure 3-11 shows an experimental encoder circuit used to convert decimal numbers 0 through 9 to binary numbers.

With power applied and all input switches open, the LED lights should all be off. This indicates that the binary number 0000 is appearing at the output. Closing decimal input switch 3, for example, inputs the decimal number 3 to the encoder. This in turn will cause the 1's and 2's LEDs to come on to indicate the equivalent binary number, 0011.

A *decoder* IC is used to convert back from *binary* to *decimal* numbers. Figure 3-12 shows an experimental decoder circuit used to convert binary numbers 0000 through 1001 back to decimal numbers. With power applied and all input switches open, the zero output LED will light, indicating binary number 0000. Closing binary input switches 2's and 4's, for example, inputs the binary number 0110 to the decoder. This in turn will cause the decimal 6 LED to come on, thus indicating the equivalent decimal number, 6.

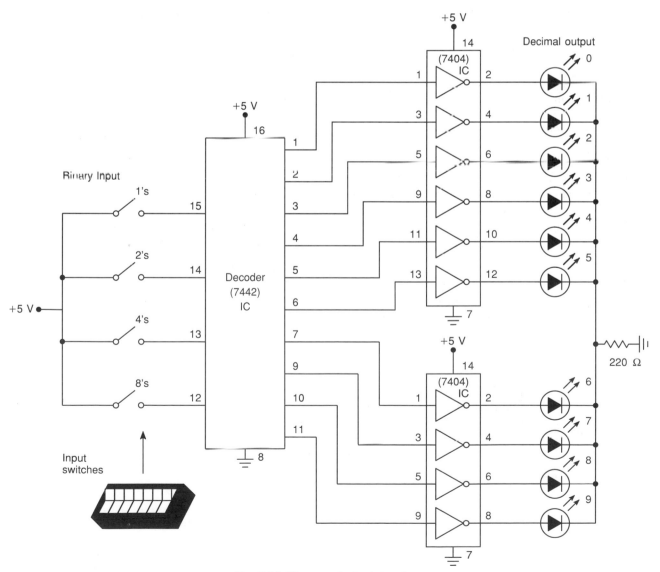

Fig. 3-12 Binary to decimal decoder.

REVIEW QUESTIONS

1. Convert each of the following binary numbers to decimal numbers:

   (a) 10
   (b) 100
   (c) 111
   (d) 1011
   (e) 1100
   (f) 10010
   (g) 10101
   (h) 11111
   (i) 11001101
   (j) 11100011

2. Convert each of the following decimal numbers to binary numbers:

   (a) 7
   (b) 19
   (c) 28
   (d) 46
   (e) 57
   (f) 86
   (g) 94
   (h) 112
   (i) 148
   (j) 230

3. Convert each of the following octal numbers to decimal numbers:

   (a) 36
   (b) 84
   (c) 120
   (d) 196
   (e) 358
   (f) 1496

4. Convert each of the following octal numbers to binary numbers:

   (a) 78
   (b) 130
   (c) 250
   (d) 348
   (e) 569
   (f) 2634

5. Convert each of the following hexadecimal numbers to decimal numbers:

   (a) 5A
   (b) C7
   (c) 9B5
   (d) 1A6

6. Convert each of the following hexadecimal numbers to binary numbers:

   (a) 4C
   (b) E8
   (c) 6D2
   (d) 31B

7. Convert each of the following decimal numbers to BCD:

   (a) 146
   (b) 389
   (c) 1678
   (d) 2502

8. What is the most important characteristic of the Gray code?

9. What is the basic function of an encoder circuit?

10. What is the basic function of a decoder circuit?

11. What makes the binary system very applicable to computer circuits?

12. Define each of the following as they apply to the binary memory locations or registers:

   (a) Bit
   (b) Byte
   (c) Word
   (d) LSB
   (e) MSB

13. State the base used for each of the following number systems:

   (a) Octal
   (b) Decimal
   (c) Binary
   (d) Hexadecimal

PROBLEMS

1. The following binary PLC sequencer code information is to be programmed using the hexadecimal code. Convert each piece of binary information to the appropriate hexadecimal code for entry into the PLC from the keyboard.

   (a) 0001 1111
   (b) 0010 0101
   (c) 0100 1110
   (d) 0011 1001

2. The encoder circuit shown in Fig. 3-13 is used to convert the decimal digits on the keyboard to a binary code. State the output status (HIGH/LOW) of A-B-C-D when decimal number

   (a) 2 is pressed
   (b) 5 is pressed
   (c) 7 is pressed
   (d) 8 is pressed

3. If the bits of a 16-bit word or register are numbered according to the octal numbering system, beginning with 00, what consecutive numbers would be used to represent each of the bits?

4. Express the decimal number 18 in *each* of the following number codes:

   (a) Binary
   (b) Octal
   (c) Hexadecimal
   (d) BCD

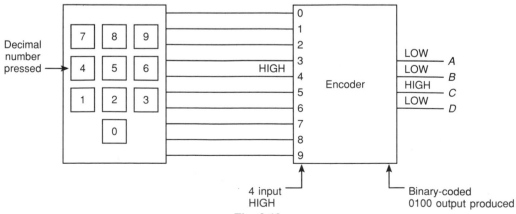

Fig. 3-13

4

FUNDAMENTALS OF LOGIC

Upon completion of this chapter you will be able to:

- Describe the binary concept and the functions of gates
- Draw the logic symbol, construct a truth table, and state the Boolean equation for the AND, OR, and NOT functions
- Construct circuits from Boolean expressions and drive Boolean equations for given logic circuits
- Convert relay ladder diagrams to logic ladder diagrams

4-1 THE BINARY CONCEPT

The PLC, like all digital equipment, operates on the binary principle. The term *binary principle* refers to the idea that many things can be thought of as existing in one of *two* states. The states can be defined as "high" or "low," "on" or "off," "yes" or "no," and "1" or "0." For instance a light can be on or off, a switch open or closed, or a motor running or stopped.

This two-state binary concept, applied to gates, can be the basis for making decisions. The gate is a device that has one or more inputs with which it will perform a logical decision and produce a result at its one output. Figures 4-1 and 4-2 give two examples that show how logic gate decisions are made.

4-2 AND, OR, AND NOT FUNCTIONS

The operations performed by digital equipment are based on three fundamental logic functions: AND, OR, and NOT. Each function has a rule that will determine the outcome and a *symbol* that represents the operation. For the purpose of this discussion, the outcome or output is called Y and the signal inputs are called A, B, C, etc. Also, binary 1 represents the presence of a signal or the occurrence of some event, while binary 0 represents the absence of the signal or nonoccurrence of the event.

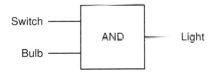

The light in a room can be turned on only when the switch is on *and* a light bulb is in the light socket.

Fig. 4-1 The logical AND.

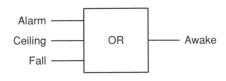

You will be awakened from your sleep when the alarm goes off *or* the ceiling caves in *or* you fall out of bed.

Fig. 4-2 The logical OR.

The AND Function

The symbol drawn in Fig. 4-3 is called an AND gate. An AND gate is a device with two or more inputs and one output. The AND gate output is 1 only if all inputs are 1. The truth table in Fig. 4-3 shows the resulting output from each of the possible input combinations.

Figures 4-4 and 4-5 show practical applications of the AND gate function.

33

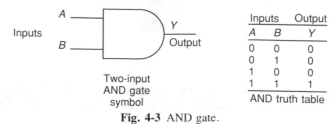

Inputs

Two-input
AND gate
symbol

Inputs		Output
A	B	Y
0	0	0
0	1	0
1	0	0
1	1	1

AND truth table

Fig. 4-3 AND gate.

All possible input combinations

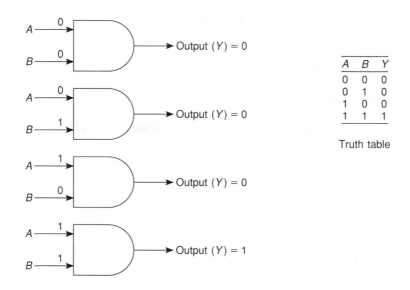

A	B	Y
0	0	0
0	1	0
1	0	0
1	1	1

Truth table

Basic rule: If all inputs are "1," the output will be "1."
If any input is "0," the output will be "0."

Fig. 4-4 AND gate function application—example 1.

Circuit

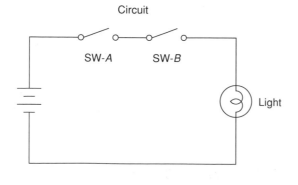

Light

SW-A		SW-B		Light	
Open	(0)	Open	(0)	Off	(0)
Open	(0)	Closed	(1)	Off	(0)
Closed	(1)	Open	(0)	Off	(0)
Closed	(1)	Closed	(1)	On	(1)

Truth table

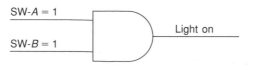

SW-A = 1

SW-B = 1

Light on

The light will be on only when both switch A and switch B are closed.

Fig. 4-5 AND gate function application—example 2.

The OR Function

The symbol drawn in Fig. 4-6 is called an OR gate. An OR gate can have any number of inputs but only one output. The OR gate output is 1 if one or more inputs are

1. The truth table in Fig. 4-6 shows the resulting output Y from each possible input combination.

Figures 4-7 and 4-8 show practical applications of the OR gate function.

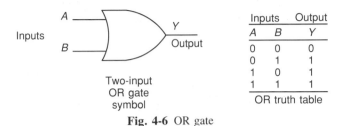

Inputs		Output
A	B	Y
0	0	0
0	1	1
1	0	1
1	1	1
OR truth table		

Two-input
OR gate
symbol

Fig. 4-6 OR gate

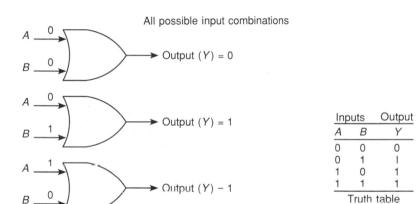

All possible input combinations

Inputs		Output
A	B	Y
0	0	0
0	1	1
1	0	1
1	1	1
Truth table		

Basic rule: If one or more inputs are "1" the output is "1."
If all inputs are "0" the output will be "0."

Fig. 4-7 OR gate function application—example 1.

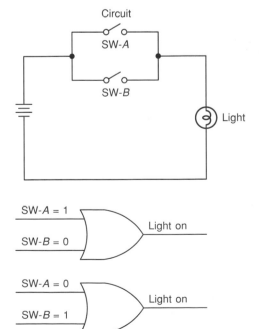

Circuit

SW-A		SW-B		Light	
Open	(0)	Open	(0)	Off	(0)
Open	(0)	Closed	(1)	On	(1)
Closed	(1)	Open	(0)	On	(1)
Closed	(1)	Closed	(1)	On	(1)
Truth table					

Light will be on if switch A or switch B is closed.

Fig. 4-8 OR gate function application—example 2.

The NOT Function

The symbol drawn in Fig. 4-9 is that of a NOT function. Unlike the AND and OR functions, the NOT function can have only *one* input. The NOT output is 1 if the input is 0. The output is 0 if the input is 1. The result of the NOT operation is always the inverse of the input and the NOT function is, therefore, called an *inverter*.

Figure 4-10 shows an example of a practical application of the NOT function.

The NOT function is most often used in conjunction with the AND or the OR gate. Figure 4-11 shows the NOT function connected to one input of an AND gate.

The NOT symbol placed at the output of an AND gate would invert the normal output result. An AND gate with

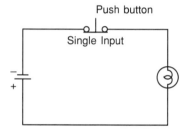

A	NOT A
0	1
1	0
NOT truth table	

Fig. 4-9 NOT function symbol

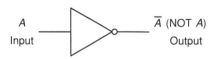

Push button	Light
Not pressed (0)	On (1)
Pressed (1)	Off (0)
Truth table	

Light will be on if the push button is not pressed.
Fig. 4-10 NOT gate function application—example 1.

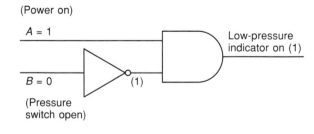

Pressure switch	Power	Pressure indicator
0	1	1
1	1	0
Truth table		

In this low-pressure-warning indicator circuit, if the power is on (1) and the pressure switch is not closed (0), the warning indicator will be on.

Fig. 4-11 NOT gate function application—example 2.

an inverted output is called a NAND gate. The NAND gate symbol and truth table are shown in Fig. 4-12.

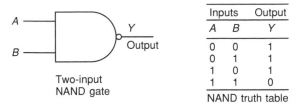

Inputs		Output
A	B	Y
0	0	1
0	1	1
1	0	1
1	1	0

NAND truth table

Fig. 4-12 NAND gate symbol and truth table.

The same rule applies if a NOT symbol is placed at the output of the OR gate. The normal output is inverted, and the function is referred to as a NOR gate. The NOR gate symbol and truth table are shown in Fig. 4-13.

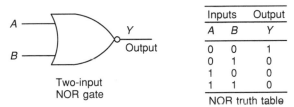

Inputs		Output
A	B	Y
0	0	1
0	1	0
1	0	0
1	1	0

NOR truth table

Fig. 4-13 NOR gate symbol and truth table

4-3 BOOLEAN ALGEBRA

The mathematical study of the binary number system and logic is called *Boolean algebra*. The purpose of this algebra is to provide a simple way of writing complicated combinations of logic statements.

Figure 4-14 summarizes the basic operators of Boolean algebra as they are related to the basic AND, OR, and NOT functions. Inputs are represented by capital letters A, B, C, etc. and the output by a capital Y. The multiplication sign ($\times$) represents the AND operation, an addition sign ($+$) represents the OR operation, and a bar over the letter $(\overline{A})$ *represents the NOT operation.*

Digital systems may be designed using Boolean algebra. Circuit functions are represented by Boolean equations. See Figs. 4-15 and 4-16 for two examples that illustrate how the basic AND, OR, and NOT functions are used to form Boolean equations.

An understanding of the technique of writing simplified Boolean equations for complex logical statements is a useful tool when creating PLC control programs. Some laws of Boolean algebra are different from those of ordinary algebra. The following three basic laws illustrate the close comparison between Boolean algebra and ordinary algebra, as well as one major difference between the two.

COMMUTATIVE LAW

$$A + B = B + A$$

$$A \cdot B = B \cdot A$$

ASSOCIATIVE LAW

$$(A + B) + C = A + (B + C)$$

$$(A \cdot B) \cdot C = A \cdot (B \cdot C)$$

DISTRIBUTIVE LAW

$$A \cdot (B + C) = (A \cdot B) + (A \cdot C)$$

$$A + (B \cdot C) = (A + B) \cdot (A + C)$$

This law holds true only in Boolean algebra.

Logic symbol	Logic statement	Boolean equation
A, B → Y	Y is "1" if A and B are "1"	$Y = A \cdot B$ or $Y = AB$
A, B → Y	Y is "1" if A or B is "1"	$Y = A + B$
A → Y	Y is "1" if A is "0" Y is "0" if A is "1"	$Y = \overline{A}$

Fig. 4-14 Boolean algebra as related to AND, OR, and NOT functions.

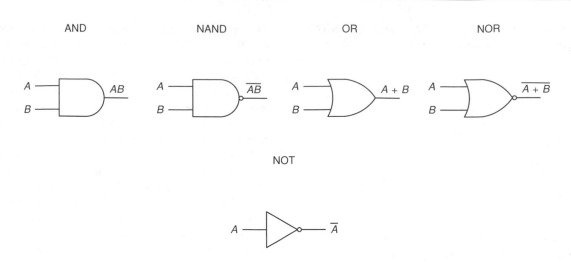

Basic logic gates inplement simple logic functions.
Each logic function can be expressed in terms of a
Boolean expression as shown.

Fig. 4-15 Boolean equation—example 1.

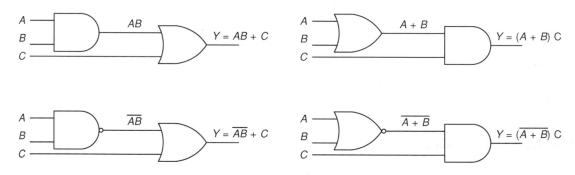

Any combination of control functions can be expressed as shown.

Fig. 4-16 Boolean equation—example 2.

De Morgan's law is one of the most important results of Boolean algebra. It shows that any logical function can be implemented with AND gates and inverters or OR gates and inverters (See Fig. 4-17).

4-4 DEVELOPING CIRCUITS FROM BOOLEAN EXPRESSIONS

Figures 4-18 and 4-19 illustrate the method used to develop a circuit from a Boolean expression.

4-5 PRODUCING THE BOOLEAN EQUATION FROM A GIVEN CIRCUIT

Figures 4-20 and 4-21 illustrate how to produce the Boolean equation from a given circuit.

4-6 HARD-WIRED LOGIC VERSUS PROGRAMMED LOGIC

The term *hard-wired logic* refers to logic control functions that are determined by the way devices are interconnected. Hard-wired logic can be implemented using relays and relay ladder diagrams. Relay ladder diagrams are universally used and understood in industry. Figure 4-22 shows a typical relay ladder diagram of a motor STOP/START control station with pilot lights. The control scheme is drawn between two vertical supply lines. All the components are placed between these two lines, called *rails* or *legs,* connecting the two power lines with what look like *rungs* of a ladder—thus the name, *ladder diagram.*

Hard-wired logic is fixed, and is changeable only by altering the way devices are connected. In contrast, programmable control is based on the basic logic functions,

According to De Morgan's laws:

$$\overline{AB} = \overline{A} + \overline{B}$$

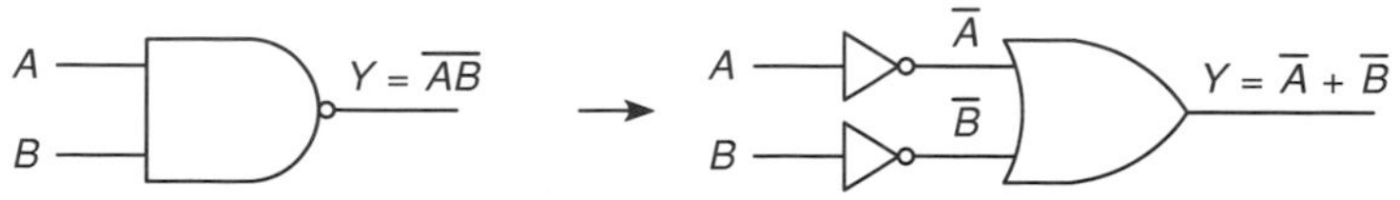

and

$$\overline{A + B} = \overline{A}\overline{B}$$

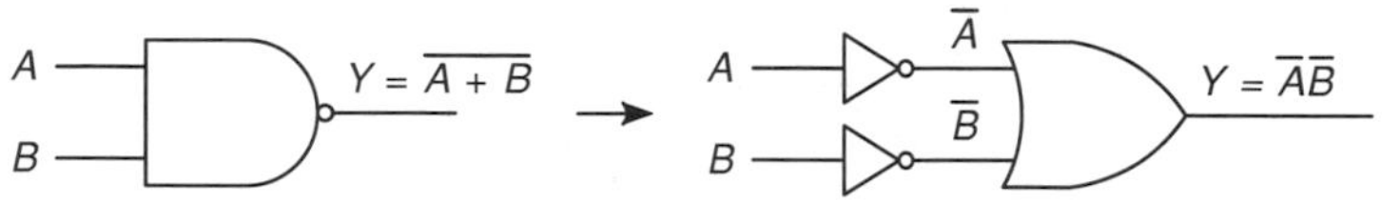

Fig. 4-17 De Morgan's laws.

Circuit diagram

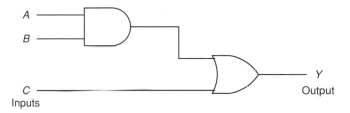

Boolean expression: $Y = AB + C$
Gates required: (by inspection)
   1   AND gate with input A and B
   1   OR gate with input C and output from previous AND gate

Fig. 4-18 Circuit development using a Boolean expression—example 1.

Circuit diagram

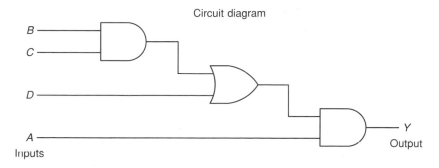

Boolean expression: $Y = A(BC + D)$
Gates required: (by inspection)
   1   AND gate with inputs B and C
   1   OR gate with inputs $B \cdot C$ *and* D
   1   AND gate with inputs A and the output
           from the OR gate

Fig. 4-19 Circuit development using a Boolean expression—example 2.

Write the Boolean equation for the following circuit:

Original circuit

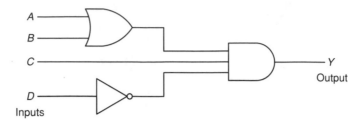

Circuit with Boolean terms

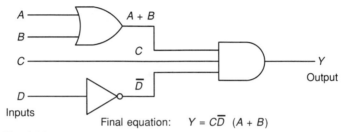

Final equation: $Y = C\overline{D} \ (A + B)$

Fig. 4-20 Producing a Boolean equation from a given circuit—example 1.

Write the Boolean equation for the following circuit:

Original circuit

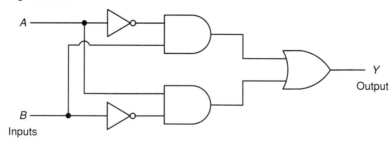

Circuit with Boolean terms

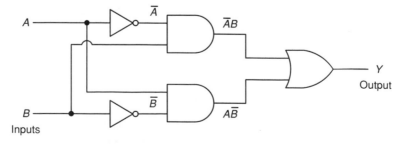

Final equation: $Y = \overline{A}B + A\overline{B}$

Fig. 4-21 Producing a Boolean equation from a given circuit—example 2.

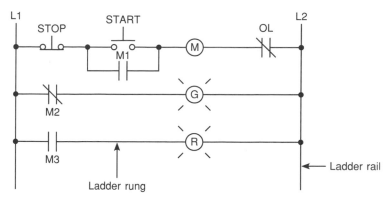

Fig. 4-22 Relay ladder diagram.

which are programmable and easily changed. These functions (AND, OR, NOT) are used either singly or in combinations to form instructions that will determine if a device is to be switched on or off. The form in which these instructions are implemented to convey commands to the PLC is called the *language*. The most common PLC language is *logic ladder diagrams*.

In Fig. 4-23 you can see a typical logic ladder diagram program for the relay ladder diagram of Fig. 4-22. The instructions used are the relay equivalent of normally open

(NO) and normally closed (NC) contacts and coils. Contact symbology is a very simple way of expressing the control logic in terms of symbols that are used on relay control schematics. A rung is the contact symbology required to control an output. Some PLCs allow a rung to have multiple outputs, but one output per rung is the convention. A complete logic ladder diagram program then consists of several rungs, each of these rungs controlling an output.

Because the PLC uses logic ladder diagrams, the con-

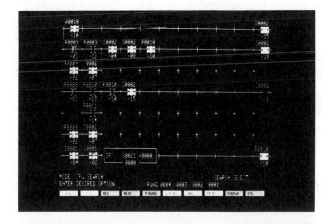

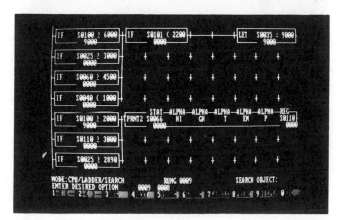

Typical PLC programming languages. *(Courtesy of Square D Company)*

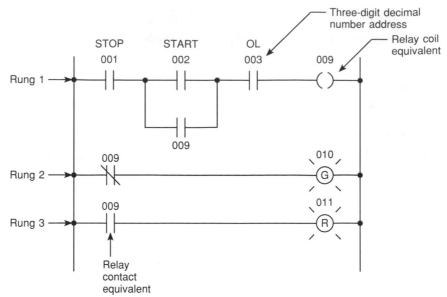

Fig. 4-23 Logic ladder diagram program.

version from any existing relay logic to programmed logic is simple. Each rung is a combination of input conditions (symbols) connected from left to right, with the symbol that represents the output at the far right. The symbols that represent the inputs are connected in series, parallel, or some combination to obtain the desired logic. The group of examples that follow illustrate the relationship between the relay ladder diagram, the logic ladder diagram program, and the equivalent logic gate circuit (See Examples 4-1 to 4-7.

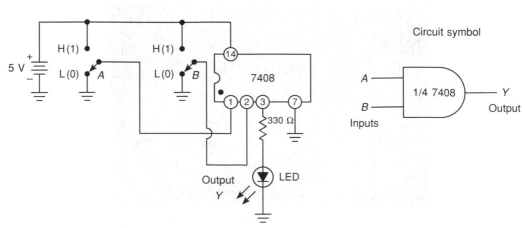

Example 4-1 Two limit switches connected in series used to control a solenoid valve.

Relay ladder diagram

Logic ladder diagram program

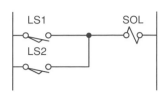

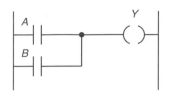

Boolean equation: $A + B = Y$

Logic gate schematic

Circuit symbol

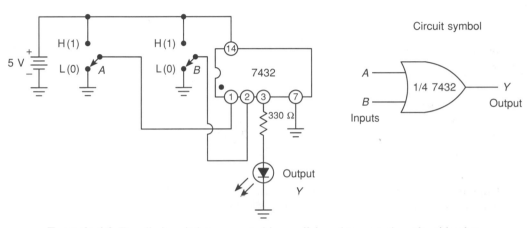

Example 4-2 Two limit switches connected in parallel used to control a solenoid valve.

Relay ladder diagram

Logic ladder diagram program

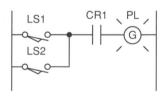

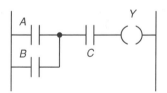

Boolean equation: $(A + B) C = Y$

Logic gate schematic

Circuit symbol

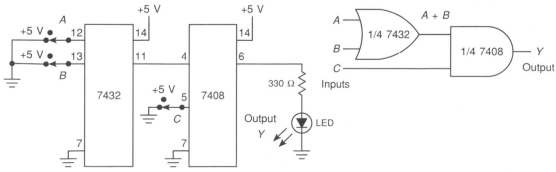

Example 4-3 Two limit switches connected in parallel with each other and in series with a relay contact used to control a pilot light.

Relay ladder diagram

Logic ladder diagram program

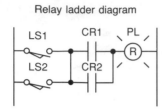

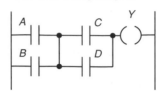

Boolean equation: $(A + B) (C + D) = Y$

Circuit symbol

Logic gate schematic

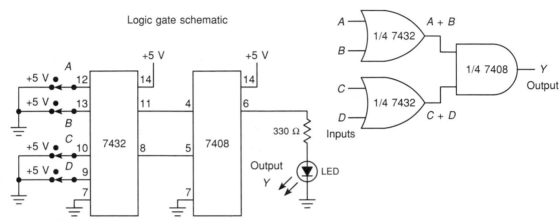

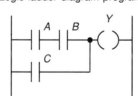

Example 4-4 Two limit switches connected in parallel with each other and in series with two sets of contacts that are connected in parallel with each other, used to control a pilot light.

Relay ladder diagram

Logic ladder diagram program

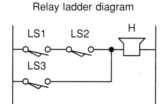

Boolean equation: $(AB) + C = Y$

Circuit symbol

Logic gate schematic

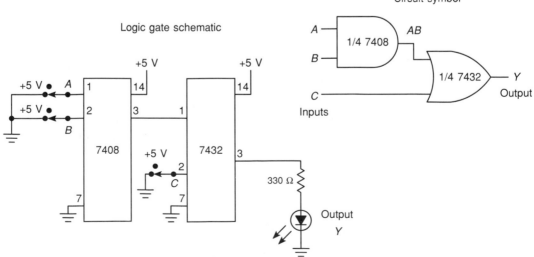

Example 4-5 Two limit switches connected in series with each other and in parallel with a third limit switch, used to control a warning horn.

Relay ladder diagram

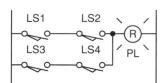

Logic ladder diagram program

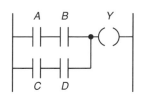

Boolean equation: $(AB) + (CD) = Y$

Circuit symbol

Logic gate schematic

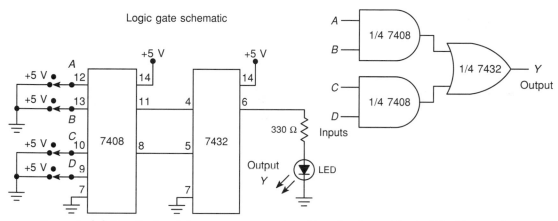

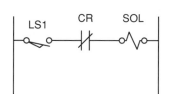

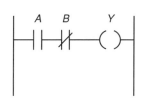

Example 4-6 Two limit switches connected in series with each other and in parallel with two other switches that are connected in series with each other, used to control a pilot light.

Relay ladder diagram

Logic ladder diagram program

Boolean equation: $A\overline{B} = Y$

Circuit symbol

Logic gate schematic

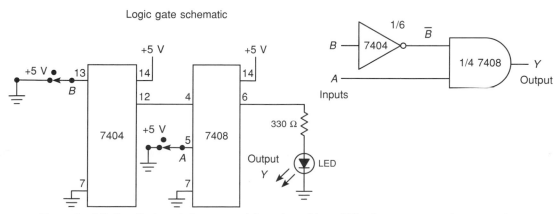

Example 4-7 One limit switch connected in series with an NC relay contact used to control a solenoid valve.

REVIEW QUESTIONS

1. Explain the binary concept.

2. What is the purpose of an electronic gate?

3. Draw the logic symbol, construct a truth table, and state the Boolean equation for each of the following:
   (a) Two-input AND gate
   (b) NOT function
   (c) Three-input OR gate

4. Express each of the following equations as a logic ladder diagram rung:

   (a) $Y = (A + B)\ CD$
   (b) $Y = A\bar{B}C + \bar{D} + E$
   (c) $Y = [(\bar{A} + \bar{B})C] + DE$
   (d) $Y = (\bar{A}B\bar{C}) + (D\bar{E}F)$

5. Draw the logic diagram and the logic gate circuit symbol, and write the Boolean equation for the following two relay ladder diagrams: (See Fig. 4-24 below).

6. Develop a circuit for each of the following Boolean expressions using AND, OR, and NOT gates:
   (a) $Y = ABC + D$
   (b) $Y = AB + CD$
   (c) $Y = (A + B)(\bar{C} + D)$
   (d) $Y = \bar{A}\ (B + CD)$
   (e) $Y = AB + C$
   (f) $Y = (ABC + D)\ (E\bar{F})$

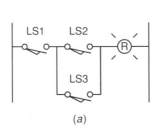

(a)

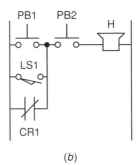

(b)

Fig. 4-24

PROBLEMS

1. It is required to have a pilot light come on when *all* of the following circuit requirements are met:

   - All four circuit pressure switches must be closed.
   - Two out of three circuit limit switches must be closed.
   - The reset switch must *not* be closed.

   Using AND, OR, and NOT gates design a logic circuit that will solve this hypothetical problem.

2. Write the Boolean equation for each of the following logic gate circuits (Fig. 4-25a-f).

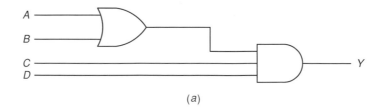

(a)

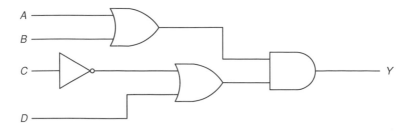

(b)

Fig. 4-25(a) and (b)

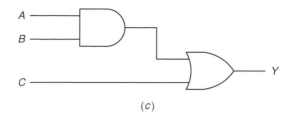

(c)

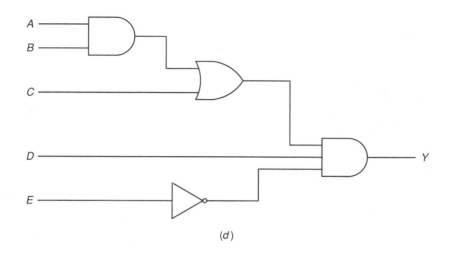

(d)

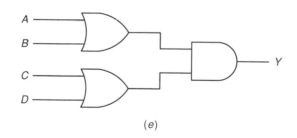

(e)

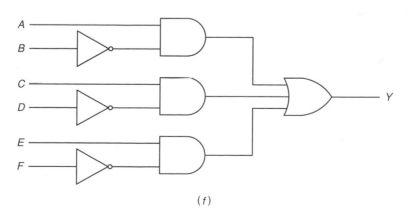

(f)

Fig. 4-25(c) to (f)

3. Match each of the following situations (i) to (v) with the analogous logic circuit (Fig. 4-26):

Logic circuits

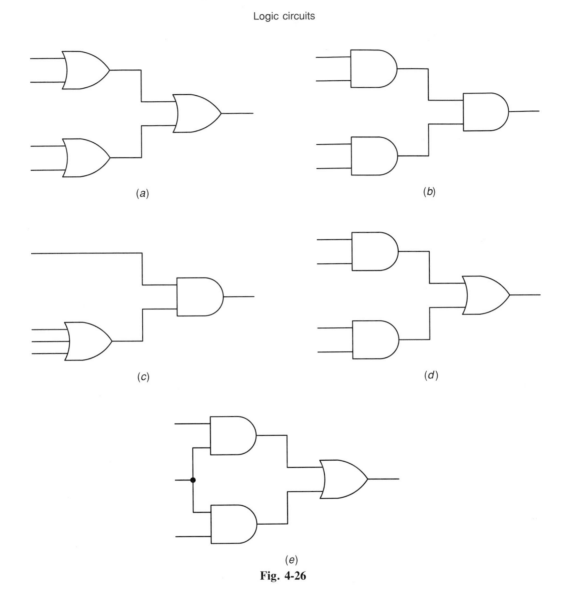

Fig. 4-26

(i)   To purchase the car, you must have $10,000.00 and a trade-in, or come up with another $2,000.00.

(ii) Two representatives from management and two from the union must be in attendance for the arbitration meeting. If a person from either group fails to show up, the meeting is called off.

(iii) To obtain a credit in the course you must be registered and also pass at least one of the major tests.

(iv) A pair of kings or a pair of aces will win the hand.

(v)   To qualify as a participant you must attend at least one morning or afternoon session of either day of the conference.

4. The following logic circuit is used to activate an alarm when its output Y is logic HIGH or 1. Draw a truth table for the circuit showing the resulting output for all 16 of the possible input conditions.

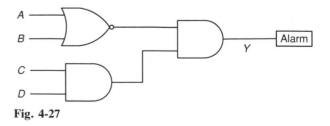

Fig. 4-27

5

BASICS OF PLC PROGRAMMING

Upon completion of this chapter you will be able to:

- Define and identify the functions of a PLC memory map
- Describe input and output image tables and a typical PLC program scan sequence
- Understand how ladder diagram language and Boolean language are used to communicate information to the PLC
- Define and identify the function of internal relay insturctions
- Identify the common operating modes found in PLCs

5-1 PROCESSOR MEMORY ORGANIZATION

The term *processor memory organization* refers to how certain areas of memory in a given PLC are used. Not all PLC manufacturers organize their memories in the same way. Although they do not all use the same memory makeup and terminology, the principles involved are the same.

Figure 5-1 shows an illustration of memory organization known as a *memory map*. Every PLC has a memory map, but it may not be like the one illustrated. The memory space can be divided into two broad categories: the *user program* and the *data table*.

The user program is where the programmed logic ladder diagram is entered and stored. The user program will account for most of the total memory of a given PLC system. It contains the logic that controls the machine operation. This logic consists of *instructions* that are programmed in a ladder logic format. Most instructions require one *word* of memory.

The data table stores the information needed to carry out the user program. This includes such information as the status of input and output devices, timer and counter values, data storage, and so on. Contents of the data table can be divided into two categories: *status data* and *numbers or codes*. Status is ON/OFF type of information represented by 1's and 0's, stored in unique bit lo-

cations. Number or code information is represented by groups of bits which are stored in unique byte or word locations.

The data table can be divided into the following three sections according to the type of information to be remembered: input image table, output image table, and timer and counter storage. The *input image table* stores the status of digital inputs, which are connected to input interface circuits. Figure 5-2 shows a typical connection of a switch to the input image table through the input module. When the switch is closed, the processor detects a voltage at the input terminal and records that information by storing a binary 1 in the proper bit location. Each connected input has a bit in the input image table that corresponds exactly to the terminal to which the input is connected. The input image table is constantly being changed to reflect the current status of the switch. If the input is on (switch closed), its corresponding bit in the table is set to 1. If the input is off (switch open), the corresponding bit is "cleared," or reset to 0.

The *output image table* is an array of bits that controls the status of digital output devices, which are connected to output interface circuits. Figure 5-3 shows a typical connection of a light to the output image table through the output module. The status of this light (ON/OFF) is controlled by the user program and indicated by the presence of 1's (ON) and 0's (OFF). Each connected output has a bit in the output image table that corresponds

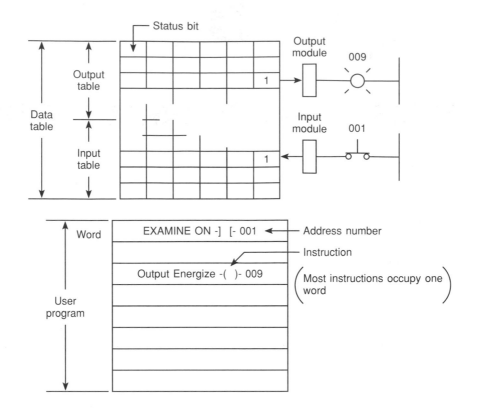

Instructions are stored in the same order you enter them. During operation, the processor carries out these instructions in this same order.

The *address number* you assign to an *instruction* associates it with a particular *status bit*. This bit will be either *on* (logic 1) or *off* (logic 0), indicating whether the instruction is *true* or *false*.

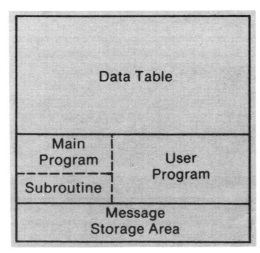

Fig. 5-1 PLC Memory map.   *(Courtesy of Allen-Bradley Company, Inc.)*

exactly to the terminal to which the output is connected. If the program calls for a specific output to be on, its corresponding bit in the table is set to 1. If the program calls for the output to be off, its corresponding bit in the table is set to 0.

5-2 PROGRAM SCAN

During each operating cycle, the processor reads all the inputs, takes these values, and according to the user program energizes or de-energizes the outputs. This process

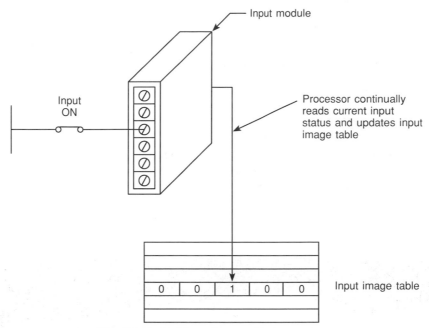

Fig. 5-2 Typical input image table connection.

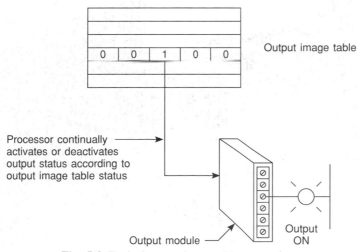

Fig. 5-3 Typical output image table connection.

is known as a *scan*. Figure 5-4 illustrates a single PLC scan, which consists of the *I/O scan* and the *program scan*.

The PLC scan time specification indicates how fast the controller can react to changes in inputs. Scan time varies with program content and length. The time required to make a single scan can vary from 1 ms to 100 ms. If a controller has to react to an input signal that changes states twice during the scan time, it is possible that the PLC will never be able to detect this change.

The scan is normally a continuous and sequential process of reading the status of inputs, evaluating the control logic, and updating the outputs. Figure 5-5 illustrates this process. When the input device connected to address 101-14 is closed, the input module circuitry senses a voltage and a 1 (ON) condition is entered into the input image table bit 101-14. During the program scan, the processor examines bit 101-14 for a 1 (ON) condition. In this case, since input 101-14 is 1, the rung is said to be TRUE. The processor then sets the output image table

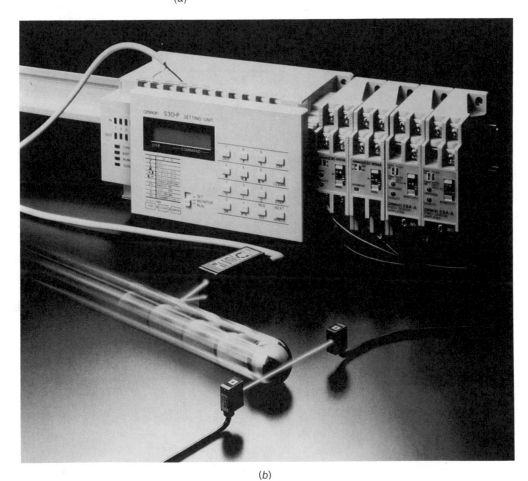

I/O scan—Records status data of input devices. Energizes output devices which have their associated status bits set to ON.

Program scan—Instructions are executed sequentially, as entered.

(a)

(b)

Fig. 5-4 (*a*) Single PLC scan. (*b*) Ideal for high-speed sensing and machine control applications, Omron's model S3D8 Sensor Controller accepts inputs at rates to 3kHz and responds in 1 millesecond.   (*Courtesy of Omron Electronics, Inc.*)

bit 001-04 to 1. The processor turns on output 001-04 during the next I/O scan, and the output device (light) wired to this terminal becomes energized. This process is repeated as long as the processor is in the RUN mode. If the input device were to open, a 0 would be placed in the input image table. As a result, the rung would be said to be FALSE. The processor would then set the output image table bit 001-04 to 0, causing the output device to turn off.

5-3 PLC PROGRAMMING LANGUAGES

The term *PLC programming language* refers to the method by which the user communicates information to the PLC. The two most common language structures are *ladder diagram language* and *Boolean language*. Although each language structure is similar from one PLC model to another, there are differences between manufacturers in the method of application. However, these

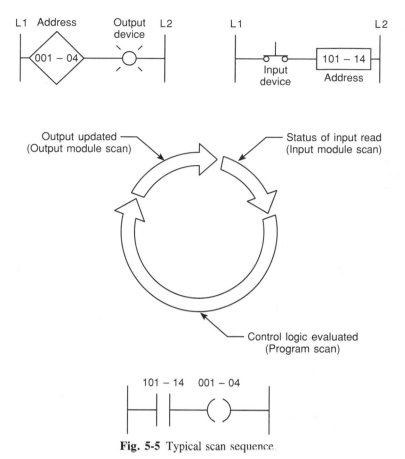

Fig. 5-5 Typical scan sequence.

differences are usually minor and easy to understand.

Ladder diagram language is by far the most commonly used PLC language. Figure 5-6 shows a comparison of ladder diagram language and Boolean language programming. Figure 5-6a shows the original relay ladder diagram drawn as if it were to be hard-wired. Figure 5-6b shows the equivalent logic ladder diagram that is programmed into the controller. Note that the addressing format shown for input and output devices is generic in nature and varies for different PLC models. Figure 5-6c shows a typical set of generic Boolean statements that could also be used to program the original circuit. This statement refers to the basic AND, OR, and NOT logic gate functions. Also included is the typical Boolean equation for the circuit.

5-4 RELAY-TYPE INSTRUCTIONS

The ladder diagram language is basically a *symbolic* set of instructions used to create the controller program. These ladder instruction symbols are arranged to obtain the desired control logic that is to be entered into the memory of the PLC. Because the instruction set is composed of contact symbols, ladder diagram language is also referred to as *contact symbology*.

Representations of contacts and coils are the basic symbols of the logic ladder diagram instruction set. The fol-

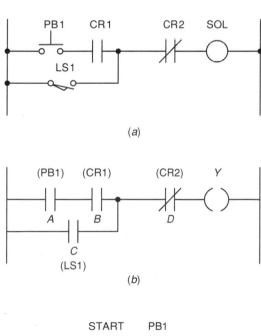

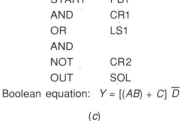

Boolean equation: $Y = [(AB) + C]\,\overline{D}$

(c)

Fig. 5-6 PLC ladder and Boolean languages.

lowing three are the basic symbols used to translate relay control logic to contact symbolic logic (see Figs. 5-7 through 5-10) for the symbols and examples:

Symbol	Description
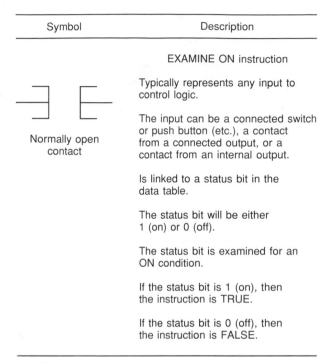 Normally open contact	**EXAMINE ON instruction**
	Typically represents any input to control logic.
	The input can be a connected switch or push button (etc.), a contact from a connected output, or a contact from an internal output.
	Is linked to a status bit in the data table.
	The status bit will be either 1 (on) or 0 (off).
	The status bit is examined for an ON condition.
	If the status bit is 1 (on), then the instruction is TRUE.
	If the status bit is 0 (off), then the instruction is FALSE.

Fig. 5-7 Normally open contact symbol.

Symbol	Description
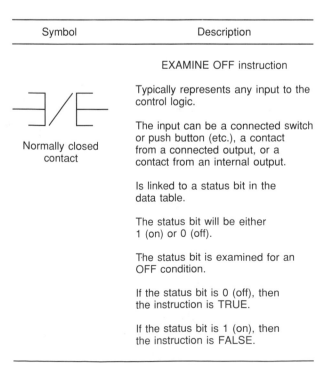 Normally closed contact	**EXAMINE OFF instruction**
	Typically represents any input to the control logic.
	The input can be a connected switch or push button (etc.), a contact from a connected output, or a contact from an internal output.
	Is linked to a status bit in the data table.
	The status bit will be either 1 (on) or 0 (off).
	The status bit is examined for an OFF condition.
	If the status bit is 0 (off), then the instruction is TRUE.
	If the status bit is 1 (on), then the instruction is FALSE.

Fig. 5-8 Normally closed contact symbol.

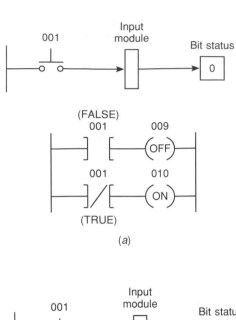

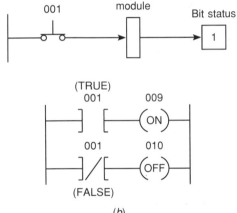

Fig. 5-9 Status bit examples. (*a*) Button not actuated. (*b*) Button actuated.

Symbol	Description
Coil	**OUTPUT ENERGIZE**
	Typically represents any output that is controlled by some combination of input logic.
	An output can be a connected device or an internal output (internal relay).
	If any left-to-right path of input conditions is TRUE, the output is energized (turned on).
	The status bit of the addressed OUTPUT ENERGIZE instruction is set to 1 (on) when the rung is TRUE.
	The status bit of the addressed OUTPUT ENERGIZE instruction is reset to 0 (off) when the rung is FALSE.

Fig. 5-10 Coil symbol.

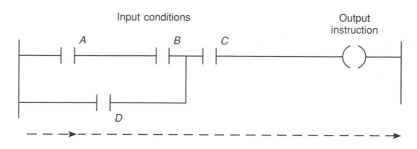

Continuous path is required for *logic continuity*, and to energize the output.

Rung condition is TRUE if contacts *A, B, C* or *D, C* are closed.

Fig. 5-11 Ladder rung.

The main function of the logic ladder diagram program is to control outputs based on input conditions. This control is accomplished through the use of what is referred to as a *ladder rung*. In general, a rung consists of a set of input conditions, represented by contact instructions, and an output instruction at the end of the rung represented by the coil symbol (Fig. 5-11). Each contact or coil symbol is referenced with an address number that identifies what is being evaluated and what is being controlled. The same contact instruction can be used throughout the program whenever that condition needs to be evaluated. For an output to be activated or energized, at least *one* left-to-right path of contacts must be closed. A complete closed path is referred to as having *logic continuity*. When logic continuity exists in at least one path, the rung condition is said to be TRUE. The rung condition is FALSE if no path has continuity.

5-5 INSTRUCTION ADDRESSING

To complete the entry of a relay-type instruction, you must assign an *address* number to it. This number will indicate what PLC input is connected to what input device and what PLC output will drive what output device.

The addressing of real inputs and outputs, as well as internals, depends upon the PLC model used. These addresses can be represented in decimal, octal, or hexadecimal depending upon the number system used by the PLC. Figure 5-12 shows a typical addressing format. Again, the programming manual of the PLC you are using should be consulted to determine the specific format used, as this can vary from model to model, as well as from manufacturer to manufacturer.

The address identifies the function of an instruction and links it to a particular *status* bit in the data table portion of the memory. Figure 5-13 shows the structure of a 16-bit word and its assigned bit values. The breakdown of this word and its addressing are in the decimal numbering system.

Format

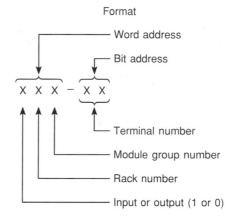

Inputs

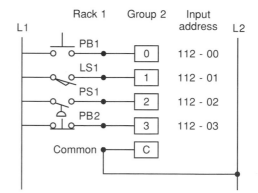

Outputs

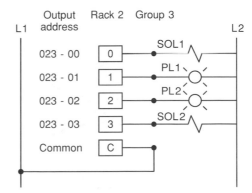

Fig. 5-12 Typical addressing format.

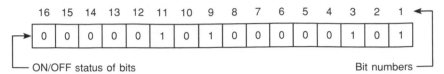

Fig. 5-13 Structure of a 16-bit word.

The assignment of I/O address is often included in the I/O connection diagram as shown in Fig. 5-14. Inputs and outputs are typically represented by squares and diamonds, respectively.

5-6 BRANCH INSTRUCTIONS

Branch instructions are used to create parallel paths of input condition instructions. This allows more than one combination of input conditions (OR logic) to establish logic continuity in a rung. Figure 5-15 illustrates a simple branching condition. The rung will be TRUE if either instruction 110/00 or 110/01 is TRUE. A branch START instruction is used to begin each parallel logic branch. A single branch CLOSE instruction is used to close the parallel branch.

In *some* PLC models the programming of a branch circuit within a branch circuit or a *nested* branch cannot be done directly. It is possible, however, to program a logically equivalent branching condition. Figure 5-16 shows an example of a circuit that contains a nested contact D. To obtain the required logic the circuit would be programmed as shown in Fig. 5-17. The duplication of contact C eliminates the nested contact D.

For each PLC model there is a limit to the number of series contact instructions that can be included in one rung of a ladder diagram, as well as a limit to the number of parallel branches. Also, there is a further limitation with some PLCs: only one output per rung, and the output must be on the first line. The only limitation on the number of rungs is memory size. Figure 5-18 shows the matrix limitation diagram for a typical PLC. A maximum of seven parallel lines and ten series contacts per rung is possible.

Another limitation to branch circuit programming is that the PLC will not allow for programming of vertical contacts. A typical example of this is contact C of the user program drawn in Fig. 5-19. To obtain the required

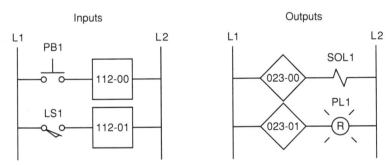

Fig. 5-14 I/O connection diagram.

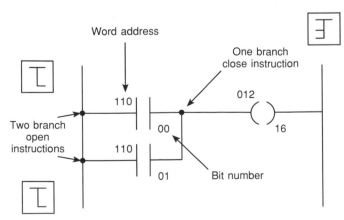

Fig. 5-15 Parallel path (branch) instructions.

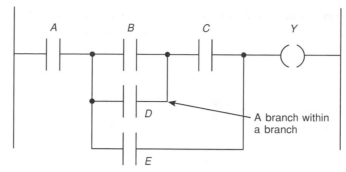

Fig. 5-16 Nested contact program.

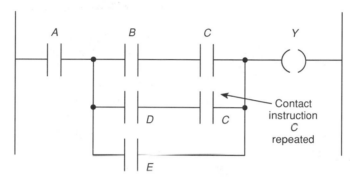

Fig. 5-17 Program required to eliminate nested contact.

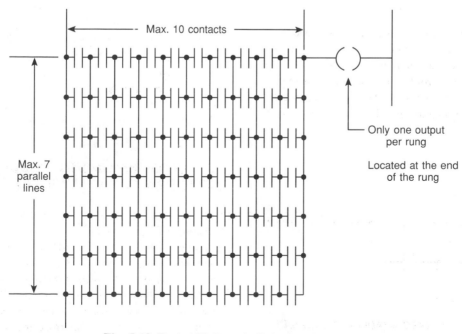

Fig. 5-18 Typical PLC matrix limitation diagram.

logic, the circuit would be reprogrammed as shown in Fig. 5-20.

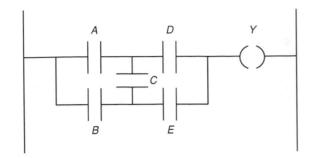

Boolean equation: $Y = (AD) + (BCD) + (BE) + (ACE)$

Fig. 5-19 Program with vertical contact.

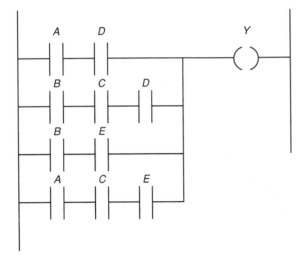

Fig. 5-20 Reprogrammed to eliminate vertical contact.

As mentioned, the processor examines the ladder logic rung for power flow from left to right *only* for logic continuity. The processor never allows for flow from right to left. This presents a problem for user program circuits similar to that shown in Fig. 5-21. If programmed as shown, contact combination *FDBC* would be ignored.

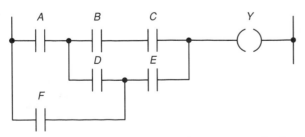

Boolean equation: $Y = (ABC) + (ADE) + (FE) + (FDBC)$

Fig. 5-21 Original circuit.

To obtain the required logic, the circuit would be reprogrammed as shown in Fig. 5-22.

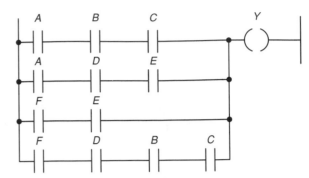

Fig. 5-22 Reprogrammed circuit.

5-7   INTERNAL RELAY INSTRUCTIONS

Most PLCs have an area of the memory allocated for what are known as *internal storage bits*. These storage bits are also called *internal outputs, internal coils, internal control relays,* or just *internals*. The internal output operates just as any output that is controlled by programmed logic; however, the output is used strictly for internal purposes. In other words, the internal output does not directly control an output device.

An internal control relay can be used when a circuit requires more series contacts than the rung allows. Figure 5-23 shows a circuit that allows for only seven series contacts when twelve are actually required for the programmed logic. To solve this problem, the contacts are split into two rungs as shown in Fig. 5-24. The first rung contains seven of the required contacts and is programmed to an internal relay. The address of the internal relay, 033 in the example, would also be the address of the first EXAMINE ON contact on the second rung. The remaining five contacts are programmed, followed by the discrete output 009. When the first seven contacts close, internal output 033 would be set to 1. This would make the examine for ON contact 033 in rung 2 TRUE. If the other six contacts in rung 2 were closed, the rung would be TRUE, and the discrete output 009 would be turned on.

5-8   PROGRAMMING EXAMINE ON AND EXAMINE OFF INSTRUCTIONS

A simple program using the EXAMINE ON instruction is shown in Fig. 5-25. This figure shows a hard-wired circuit and a user program that provides the same results. You will note that *both the NO and the NC* push buttons are represented by the EXAMINE ON symbol. This is because the normal state of an input (NO or NC) does not matter to the controller. What does matter is that if contacts need to *close* to energize the output, then the EXAMINE ON instruction is used. Since both PB1 and

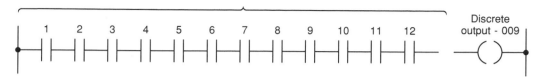

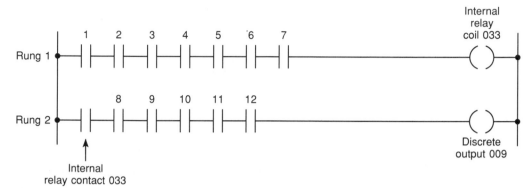

Fig. 5-23 Contacts exceed limit of seven.

Fig. 5-24 Contacts split into two rungs.

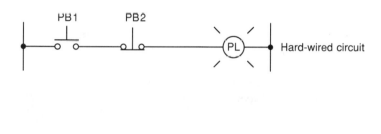

Fig. 5-25 EXAMINE ON instruction.

PB2 in Fig. 5-25 must be closed to energize the pilot light, the EXAMINE ON instruction is used for both.

A simple program using the EXAMINE OFF instruction is shown in Fig. 5-26. Again, both the hard-wired circuit and user program are shown. In the hard-wired circuit, when the push button is *open*, relay coil CR is de-energized and contacts CR1 close to switch the pilot light *on*. When the push button is *closed*, relay coil CR is energized, and contacts CR1 open to switch the pilot light *off*. The push button is represented in the user program by an EXAMINE OFF instruction. This is because the rung must be TRUE when the external push button is open, and FALSE when the push button is closed. Using an EXAMINE OFF instruction to represent the push button satisfies these requirements. The NO or NC mechan-

ical action of the push button is not a consideration. It is important to remember that the user program is not an electrical circuit but a *logic* circuit. In effect we are interested in logic continuity when establishing an output.

5-9   ENTERING THE LADDER DIAGRAM

The entering of the ladder diagram or actual programming is usually accomplished using a desktop or hand-held programming device. Because hardware and programming techniques vary with each manufacturer, it is necessary to refer to the programming manual for a specific PLC to determine how the instructions are entered. Regardless of the type of programming device used,

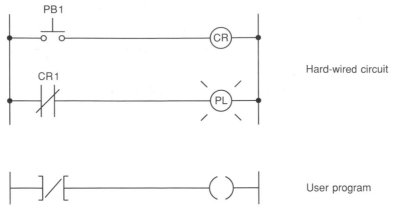

Fig. 5-26 EXAMINE OFF instruction.

some common relay symbols are standard. Three examples are shown in Fig. 5-27.

Fig. 5-27 Common relay symbols.

The method of entering a program is through the operator terminal keyboard. Keyboards usually have relay symbol and special function keys, along with numeric keys for addressing. Some also have alphanumeric (letters and numbers) keys for other special programming functions. Figure 5-28 shows a typical keyboard layout.

The ladder diagram is entered by pushing keys on the keyboard in a prescribed sequence. The results are displayed on either the CRT of a desktop programmer or with an LED or LCD for a hand-held programmer. The contacts and coils of the visual display are either intensified or displayed in reverse video to indicate the status of the instruction (Figure 5-29).

The programming manual for the PLC that you are using should be consulted before attempting any programming. The proper keystroke sequence is important for entering the program correctly into the controller and is most easily learned through hands-on experience. Figure 5-30 shows you how a typical ladder rung is programmed on the Allen-Bradley 1742 Modular Automation Controller.

For most PLC systems each EXAMINE ON and EXAMINE OFF contact, each output, and each branch START/END instruction requires one word of user memory. When the key for an EXAMINE ON contact symbol is pushed and then followed by an address, one full word of user memory is used. Many controllers will dis-

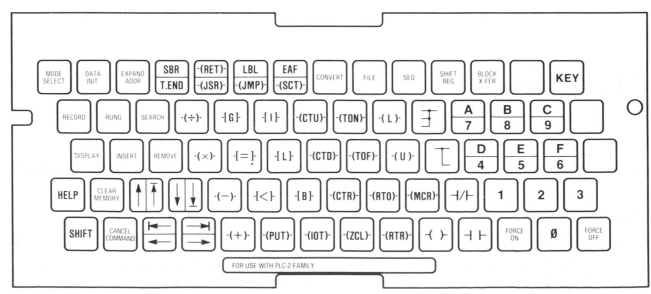

Fig. 5-28 PLC keyboard layout. *(Courtesy of Allen-Bradley Company, Inc.)*

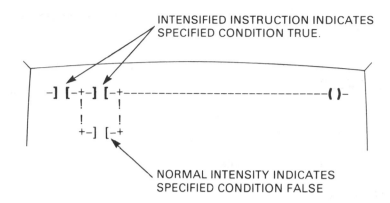

INTENSIFIED INSTRUCTION INDICATES
SPECIFIED CONDITION TRUE.

NORMAL INTENSITY INDICATES
SPECIFIED CONDITION FALSE

TRUE instructions in the rung are displayed in reverse
video, represented here by shading. The remaining instructions in
the rung are FALSE.

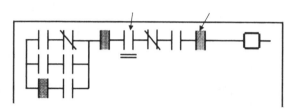

Fig. 5-29 Typical programmer display. *(Courtesy of Allen-Bradley Company, Inc.)*

Keystroke Example – Typical Ladder Rung

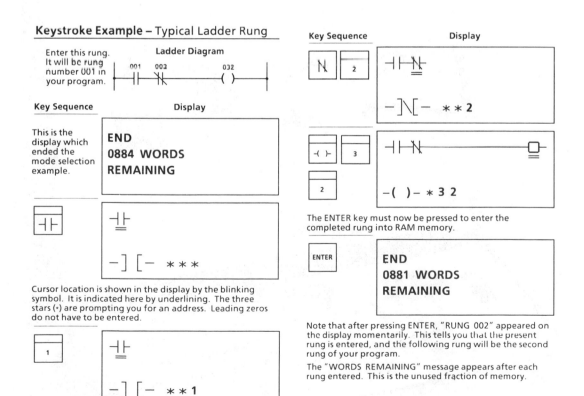

Enter this rung. It will be rung number 001 in your program.

Ladder Diagram

This is the display which ended the mode selection example.

END
0884 WORDS
REMAINING

Cursor location is shown in the display by the blinking symbol. It is indicated here by underlining. The three stars (∗) are prompting you for an address. Leading zeros do not have to be entered.

The address has been entered. If you entered an incorrect address, you can change it at this time by repeatedly pressing the "zero" key until "000" appears, then enter the correct address. Editing keys could also be used, as you will learn later.

The next instruction can now be entered.

The ENTER key must now be pressed to enter the completed rung into RAM memory.

END
0881 WORDS
REMAINING

Note that after pressing ENTER, "RUNG 002" appeared on the display momentarily. This tells you that the present rung is entered, and the following rung will be the second rung of your program.

The "WORDS REMAINING" message appears after each rung entered. This is the unused fraction of memory.

Fig. 5-30 Programming of a typical ladder rung. *(Courtesy of Allen-Bradley Company, Inc.)*

play the total number of memory words that have been used on the video display.

5-10 MODES OF OPERATION

The programming device can also be used to select the various processor modes of operation. Again, the number of different operating modes and the method of accessing them varies with the manufacturer. Regardless of PLC model, some common operating modes are CLEAR MEMORY, PROGRAM, TEST, and RUN. A chart describing the purpose of these standard modes of operation is shown in Table 5-1.

Table 5-1 PLC MODES OF OPERATION

Mode	Description
CLEAR MEMORY	Used to erase the contents of the on-board RAM memory.
PROGRAM	Used to enter a new program or update an existing one in the internal RAM memory.
TEST	Used to operate or monitor the user program without energizing any outputs.
RUN	Used to execute the user program. Input devices are monitored and output devices are energized accordingly.

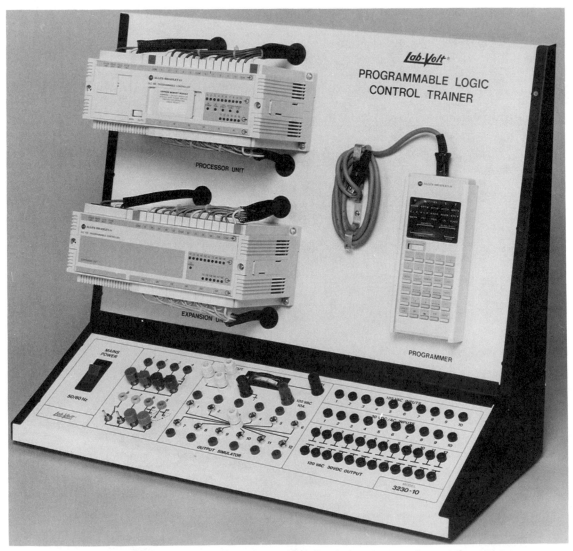

Typical programmable logic controller trainer used to develop competence in operating, programming, and trouble-shooting PLC circuits. *(Courtesy of Quintech Div. Lab-Volt (Quebec) Ltd.)*

REVIEW QUESTIONS

1. Briefly explain the purpose of the user program portion of a typical PLC memory map.

2. Briefly explain the purpose of the data table portion of a typical PLC memory map.

3. (a) What information is stored in the input image table?
   (b) In what form is this information stored?

4. (a) What information is stored in the output image table?
   (b) In what form is this information stored?

5. Outline the sequence of events involved in a single PLC program scan.

6. Draw the equivalent logic ladder diagram and write the Boolean statement and equation for the relay ladder diagram drawn here (Fig. 5-31).

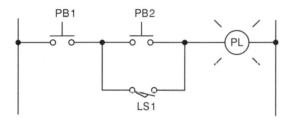

Fig. 5-31

7. Draw the symbol and state the equivalent instruction for each of the following: NO contact, NC contact, and coil.

8. (a) What does an EXAMINE ON or EXAMINE OFF instruction represent?
   (b) What does an OUTPUT ENERGIZE instruction represent?
   (c) The status bit of an EXAMINE ON instruction is examined and found to be 0. What does this mean?
   (d) The status bit of an EXAMINE OFF instruction is examined and found to be 1. What does this mean?
   (e) Under what condition would the status bit of an OUTPUT ENERGIZE instruction be 0?

9. (a) Describe the basic makeup of a logic ladder rung.
   (b) How are the contacts and coil of a rung identified?
   (c) When is the ladder rung considered TRUE, or as having logic continuity?

10. What two addresses are contained in some five-digit PLC addressing formats?

11. What is the function of an internal control relay?

12. An NO limit switch is to be programmed to control a solenoid. What determines whether an EXAMINE ON or EXAMINE OFF contact instruction is used?

13. Briefly describe each of the following modes of operation of PLCs:
   (a) CLEAR MEMORY   (c) TEST
   (b) PROGRAM   (d) RUN

PROBLEMS

1. Assign each of the inputs and outputs shown in the table the correct address based on a typical five-digit addressing.

Inputs			
Device	Terminal Number	Rack Number	Module Group Number
(a) Limit switch	1	1	3
(b) Pressure switch	2	1	3
(c) Push button	3	1	3

Outputs			
Device	Terminal Number	Rack Number	Module Group Number
(d) Pilot light	10	1	0
(e) Motor starter	11	1	0
(f) Solenoid	12	1	0

2. Draw the equivalent logic ladder diagram and write the Boolean statement and equation for the relay ladder diagram (Fig. 5-32).

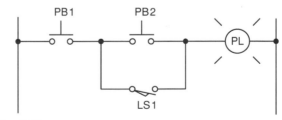

Fig. 5-32

3. Redraw each of the following programs corrected for the problem indicated:
   (a) *Problem:* nested programmed contact (Fig. 5-33).
   (b) *Problem:* vertical programmed contact (Fig. 5-34).
   (c) *Problem:* some logic ignored (Fig. 5-35).
   (d) *Problem:* too many series contacts (only four allowed) (Fig. 5-36).

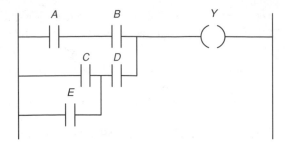

(*a*) Problem: nested programmed contact

Fig. 5-33.

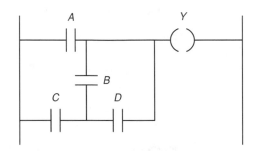

(*b*) Problem: vertical programmed contact

Fig. 5-34.

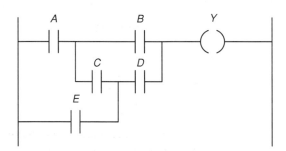

(*c*) Problem: some logic ignored

Fig. 5-35.

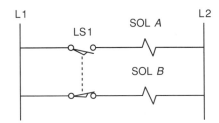

(*d*) Problem: too many series contacts (only four allowed)

Fig. 5-36.

4. Draw the equivalent ladder logic diagram used to implement the hard-wired circuit drawn here (Fig. 5-37), wired using:
   (a) A limit switch with a single NO contact connected to the PLC input module
   (b) A limit switch with a single NC contact connected to the PLC input module

Fig. 5-37

5. Assuming the hard-wired circuit drawn here (Fig. 5-38) is to be implemented using a PLC program, identify
   (a) All input field devices
   (b) All output field devices
   (c) All devices that could be programmed using internal relay instructions

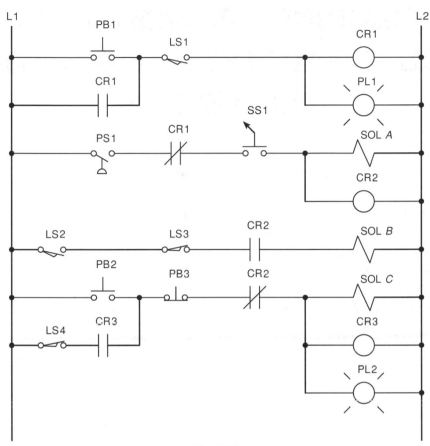

Fig. 5-38.

6

DEVELOPING FUNDAMENTAL PLC WIRING DIAGRAMS AND LADDER PROGRAMS

Upon completion of this chapter you will be able to:

- Identify the functions of electromagnetic control relays
- Identify switches commonly found in PLCs
- Describe the operation of an electromagnetic latching relay and the PLC-programmed LATCH/UNLATCH instruction
- Compare sequential and combination control processes
- Convert fundamental relay ladder diagrams to PLC logic ladder programs

6-1 ELECTROMAGNETIC CONTROL RELAYS

As previously stated, the PLC's original intent was the replacement of electromagnetic relays with a solid-state switching system that could be programmed. Although the PLC has replaced much of the relay control logic, electromagnetic relays are still used as auxiliary devices to switch I/O field devices. In addition, an understanding of electromagnetic relay operation and terminology is important for correctly converting relay ladder diagrams to logic ladder diagrams.

An electrical relay is a magnetic switch. It uses electromagnetism to switch contacts. A relay will usually have only one coil, but may have any number of different contacts. Figure 6-1 illustrates the operation of a typical control relay. With no current flow through the coil (de-energized), the armature is held away from the core of the coil by spring tension. When the coil is energized, it produces an electromagnetic field. Action of this field, in turn, causes the physical movement of the armature. Movement of the armature causes the contact points of the relay to open and close alternately. The coil and contacts are insulated from each other, therefore under nor-

mal conditions no electric circuit will exist between them.

The symbol used to represent a control relay is shown in Fig. 6-2. The contacts are represented by a pair of short parallel lines and are identified with the coil by means of the same number and letters (ICR). Both an NO and an NC contact are shown. *Normally open contacts* are defined as those contacts that are *open* when no current flows through the coil but *close* as soon as the coil conducts a current or is energized. *Normally closed contacts* are those contacts that are *closed* when the coil is de-energized and *open* when the coil is energized. Each contact is usually drawn as it would appear with the coil de-energized.

A typical control relay used to control two pilot lights is shown in Fig. 6-3. With the switch *open,* coil ICR is de-energized. The circuit to the green pilot light is completed through NC contact ICR2, so this light will be on. At the same time the circuit to the red pilot light is opened through NO contact ICR1, so this light will be off.

With the switch closed (Fig. 6-4) the coil is energized. The NO contact ICR1 closes to switch the red pilot light on. At the same time the NC contact ICR2 opens to switch the green pilot light off.

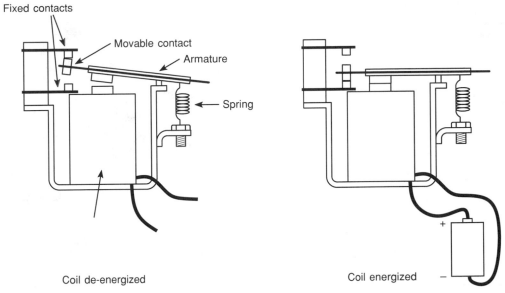

Fixed contacts

Movable contact

Armature

Spring

Coil de-energized

Coil energized

Fig. 6-1 Electromagnetic control relay operation.

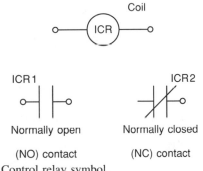

Coil

ICR

ICR1

Normally open

(NO) contact

ICR2

Normally closed

(NC) contact

Fig. 6-2 Control relay symbol.

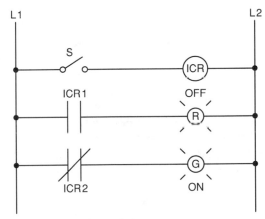

Fig. 6-3 Relay circuit—switch open.

Typical industrial control relay. *(Courtesy of Allen-Bradley Company, Inc.)*

6-2 MOTOR STARTERS

Magnetic motor starters are electromagnetically operated switches that provide a safe method for starting large motor loads. Figure 6-5 shows the wiring diagram for a

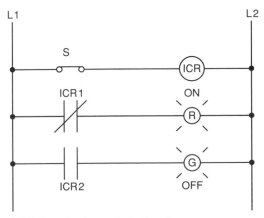

Fig. 6-4 Relay circuit—switch closed.

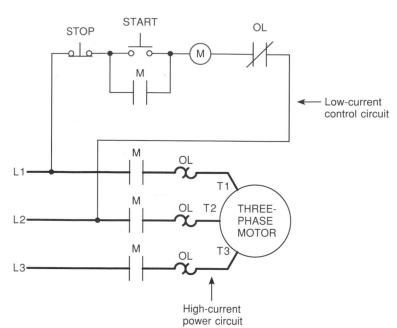

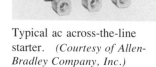

Fig. 6-5 Across-the-line ac starter.

Typical ac across-the-line starter. *(Courtesy of Allen-Bradley Company, Inc.)*

typical three-phase magnetically operated across-the-line ac starter. When the START button is pressed, coil M is energized. When coil M is energized, it closes *all* M contacts. The M contacts in series with the motor close to complete the current path to the motor. These contacts are part of the *power* circuit and must be designed to handle the full load current of the motor. Control contact M (across START button) closes to seal in the coil circuit when the START button is released. This contact is part of the *control* circuit and as such is required to handle the small amount of current needed to energize the coil. An overload (OL) relay is provided to protect the motor against current overloads. Normally closed relay contact OL opens automatically when an overload current is sensed, to de-energize M coil and stop the motor.

6-3 MANUALLY OPERATED SWITCHES

A *manually operated switch* is one that is controlled by hand. These include toggle switches, push-button switches, knife switches, and selector switches.

Three commonly used push-button switches are illustrated by their symbols in Fig. 6-6. The NO push button makes a circuit when it is pressed and returns to its open position when the button is released. The NC push button opens the circuit when it is pressed and returns to the closed position when the button is released. The break-make push button is used for *interlocking* controls. In this switch the top section is NC, while the bottom section is NO. When the button is pressed, the bottom contacts are closed as the top contacts open.

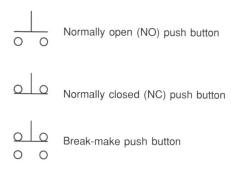

Note: The abbreviations NO and NC represent the electrical state of the switch contacts when the switch is not actuated.

Fig. 6-6 Push-button switches.

Typical pushbutton control station. *(Courtesy of Allen-Bradley Company, Inc.)*

The *selector switch* is another common manually operated switch. Selector switch positions are made by turning the operator knob—not pushing it. Selector switches may have two or more selector positions with either maintained contact position or spring return to give momentary contact operation. Figure 6-7 shows a layout of a typical control station that is used for inputs to and outputs from a PLC. This station contains four illuminated NO push buttons and four illuminated selector switches. The push buttons and selector switches serve as external input devices, while the associated pilot lights serve *independently* as external output devices. Also indicated are the external I/O addresses these devices are associated with for this particular controller.

6-4 MECHANICALLY OPERATED AND PROXIMITY SWITCHES

A *mechanically operated switch* is one that is controlled automatically by such factors as pressure, position, or temperature. The *limit switch,* shown in Fig. 6-8, is a very common industrial control device. Limit switches are designed to operate only when a predetermined limit is reached, and they are usually actuated by contact with an object such as a cam. These devices take the place of a human operator. They are often used in the control circuits of machine processes to govern the starting, stopping, or reversal of motors.

The *temperature switch,* or *thermostat,* shown in

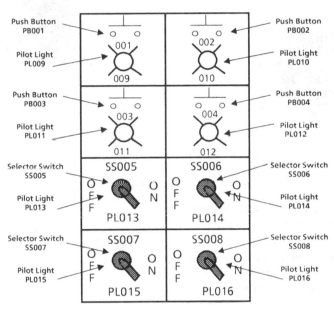

Control station I/O devices.

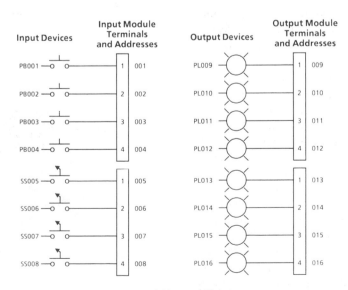

Addresses of I/O devices.

Fig. 6-7 PLC control station.

Symbols

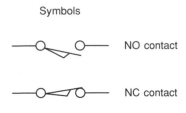

NO contact

NC contact

Fig. 6-8 Limit switches.   *(Photo courtesy of EATON Corporation, Cutler-Hammer Products)*

Fig. 6-9 is used to sense temperature changes. Although there are many types available, they are all actuated by some specific environmental temperature change. Temperature switches open or close when a designated temperature is reached. Industrial applications for these devices include maintaining the desired temperature range of air, gasses, liquids, or solids.

The *pressure switch* shown in Fig. 6-10 is used to control the pressure of liquids and gases. Again, although many types are available, they are all basically designed to actuate (open or close) their contacts when the specified pressure is reached.

Level switches, such as the one illustrated in Fig. 6-11, are used to sense the height of a liquid. The raising or lowering of a float which is mechanically attached to the level switch trips the level switch; the level switch itself is used to control motor-driven pumps that empty or fill tanks. Level switches are also used to open or close piping solenoid valves to control fluids.

A newer type of sensor switch that is becoming increasingly more popular is the *proximity switch*. Proximity switches are part of a series of solid-state sensors. They sense the presence or absence of a target *without physical contact*. The six basic types are magnetic, capacitive, ultrasonic, inductive, air jet stream, and photo electric. Figure 6-12 illustrates typical examples of industrial processes that use proximity switch sensors. The symbols for these switches are usually the same as those used for limit switches.

Symbols

NO contact

NC contact

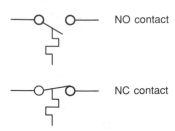

Fig. 6-9 Temperature switch.   *(Photo courtesy of Allen-Bradley Company, Inc.)*

Symbols

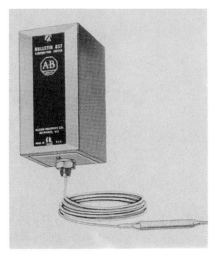

NO contact

NC contact

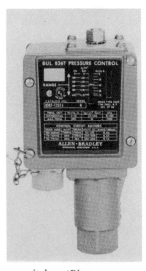

Fig. 6-10 Pressure switch.   *(Photo courtesy of Allen-Bradley Company, Inc.)*

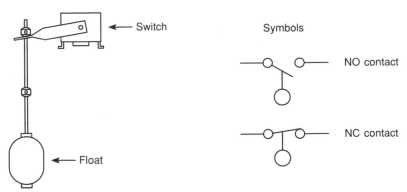

NO contact

NC contact

Fig. 6-11 Level switch.

6-5 OUTPUT CONTROL DEVICES

A variety of output control devices can be operated by the controller output module to control traditional industrial processes. These devices include pilot lights, control relays, motor starters, alarms, heaters, solenoids, solenoid valves, small motors, and horns. Electrical symbols are used to represent these devices both on relay schematic diagrams and PLC output connection diagrams. For this reason, recognition of the symbols used is important. Figure 6-13 shows common electrical symbols used for various output devices. While these symbols are

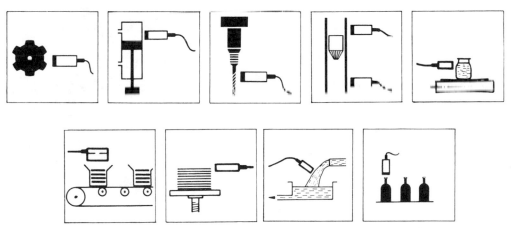

Fig. 6-12 Proximity switch applications. *(Courtesy of Rechmer Electronics Industries, Inc.)*

Typical proximity limit switch designed for industrial environments in applications where it is required to sense the presence of metal objects without touching them. *(Courtesy of Allen-Bradley Company, Inc.)*

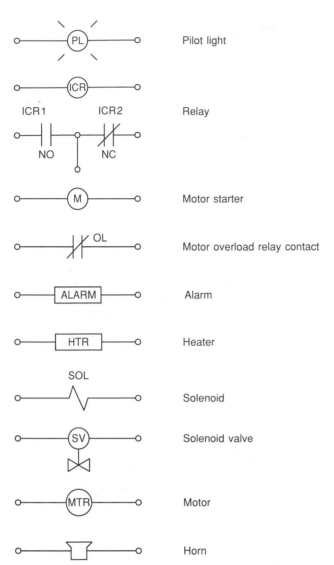

Pilot light

Relay

ICR1   ICR2

NO   NC

Motor starter

Motor overload relay contact

Alarm

Heater

Solenoid

Solenoid valve

Motor

Horn

Fig. 6-13 Symbols for output control devices.

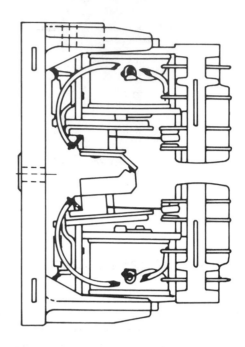

Fig. 6-14 Electromagnetic latching relay. *(Courtesy of Potter and Brumfield, Inc.)*

generally accepted by industry personnel, some differences among manufacturers do exist.

6-6 LATCHING RELAYS

Electromagnetic latching relays are designed to hold the relay closed after power has been removed from the coil. Latching relays are used where it is necessary for contacts to stay open and/or closed even though the coil is energized only momentarily. Figure 6-14 shows a latching relay that uses two coils. The *latch* coil is momentarily energized to set the latch and hold the relay in the latched position. The *unlatch* or release coil is momentarily energized to disengage the mechanical latch and return the relay to the unlatched position.

Figure 6-15 shows the schematic diagram for an electromagnetic latching-type relay. The contact is shown with the relay in the *unlatched* position. In this state the circuit to the pilot light is open and so the light is off. When the ON button is *momentarily* actuated, the latch coil is energized to set the relay to its latched position. The contacts close, completing the circuit to the pilot light and so the light is switched on. Note that the relay coil *does not* have to be continuously energized to hold the contacts closed and keep the light on. The only way to switch the lamp off is to actuate the OFF button, which will energize the unlatch coil and return the contacts to their open, unlatched state. In cases of power loss, the relay will remain in its original latched or unlatched state when power is restored.

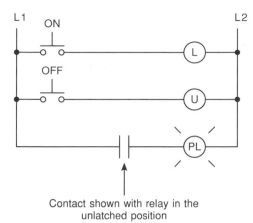

Fig. 6-15 Schematic of electromagnetic latching relay.

An electromagnetic latching relay function can be programmed on a PLC to work like its real-world counterparts. The use of the LATCH and UNLATCH coil instruction is illustrated in the ladder program of Fig. 6-16. Both the latch (L) and the unlatch (U) coil have the *same* address (015). When the ON button (001) is momentarily actuated, the latch rung becomes TRUE and the latch status bit (015) is set to ON, and so the light is switched on. This status bit *will remain on* when logical continuity of the latch rung is lost. When the unlatch

rung becomes TRUE (button 002 actuated), the status bit (015) is set back to OFF and so the light is switched off.

The LATCH/UNLATCH instruction is *retentive* on power lost. That is to say, if the relay is latched, it will remain latched if power is lost and then restored.

6-7 CONVERTING RELAY LADDER DIAGRAMS INTO PLC LADDER PROGRAMS

The best approach to developing a PLC program from a relay ladder diagram is to understand first the operation of each relay ladder rung. As each relay ladder rung is understood, an equivalent PLC rung can be generated. This will require access to the relay ladder schematic, documentation of the various input and output devices used, and possibly a process flow diagram of the operation.

Most industrial processes require the completion of several operations to produce the required output. Manufacturing, machining, assembling, packaging, finishing, or transporting of products requires the precise coordination of tasks. The majority of industrial control processors use *sequential controls*. Sequential controls are

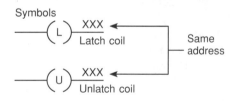

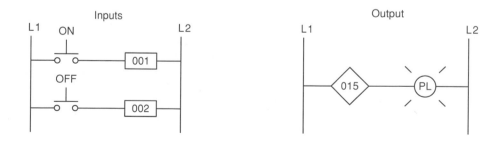

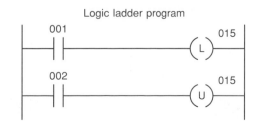

Fig. 6-16 LATCH/UNLATCH instruction program.

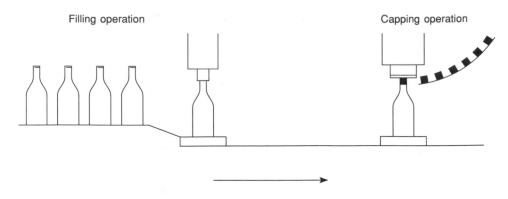

Filling operation   Capping operation

Bottle movement

Fig. 6-17 Sequential control process.

required for processes that demand that certain operations be performed in a specific order. Figure 6-17 illustrates part of a pop-bottling process. In the filling and capping operation, the tasks are (1) fill bottle, (2) press on cap. These tasks must be performed in the proper order. Obviously we could not fill the bottle after the cap was pressed on. This process, therefore, requires sequential control.

Combination controls require that certain operations be performed without regard to the order in which they are performed. Figure 6-18 illustrates another part of the same pop-bottling process. Here, the tasks are (1) place label 1 on bottle; (2) place label 2 on bottle. The order in which the tasks are performed does not really matter. In fact, however, many industrial processes that are not inherently sequential in nature are performed in a sequential manner for the most efficient order of operations.

The converting of a simple sequential process can be examined with reference to the simple task illustrated in Fig. 6-19. Shown is a process flow diagram along with the relay ladder diagram of its electrical control circuit.

The sequential task is as follows:

1. START button is pressed.
2. Table motor is started.
3. Package moves to the position of the limit switch and stops.

Other auxiliary features include:

1. An emergency STOP button that will stop the table, for any reason, before the package reaches the limit switch position
2. A red pilot light to indicate the table is stopped
3. A green pilot light to indicate the table is running

A summary of the control task for the process illustrated in Fig. 6-19 could be written as follows:

1. START button is actuated; ICR is energized if emergency STOP button and limit switch are not actuated.
2. Contact ICR1 closes, sealing in ICR even if the START button is released.
3. Contact ICR2 opens, switching the red pilot light from ON to OFF.
4. Contact ICR3 closes, switching the green pilot light from OFF to ON.
5. Contact ICR4 closes to energize the motor starter coil, starting the motor and moving the package toward the limit switch.
6. Limit switch is actuated, de-energizing relay coil ICR.
7. Contact ICR1 opens, opening the seal-in circuit.
8. Contact ICR2 closes, switching the red pilot light from OFF to ON.
9. Contact ICR3 opens, switching the green pilot light from ON to OFF.
10. Contact ICR4 opens, de-energizing the motor starter coil to stop the motor and end the sequence.

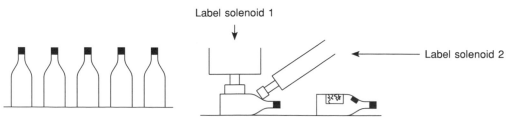

Label solenoid 1

Label solenoid 2

Fig. 6-18 Combination control process.

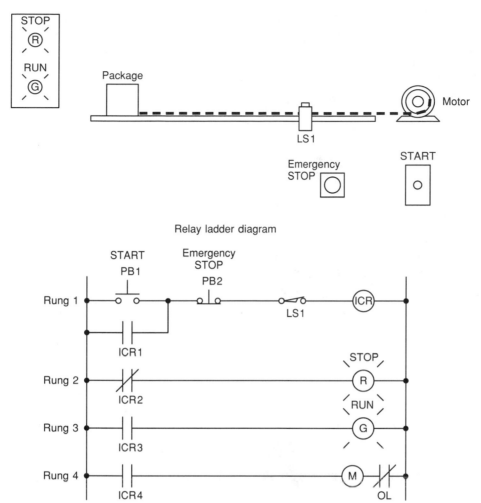

Fig. 6-19 Process flow diagram.

At this point it is wise to make an I/O connection diagram for the circuit. Each input and output device should be represented along with its address. These addresses will indicate what PLC input is connected to what input device and what PLC output will drive what output device. The address code, of course, will depend on the PLC model used. Figure 6-20 shows a typical I/O connection diagram for the process. A three-digit decimal

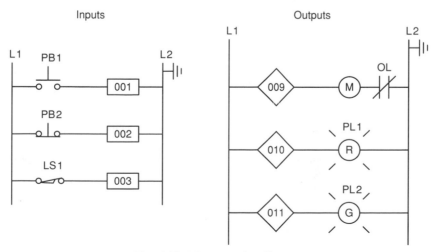

Fig. 6-20 I/O connection diagram.

numbering system is used for this example. Note that the electromagnetic control relay ICR is *not* needed, as its function is replaced by an *internal* PLC control relay.

The four rungs of the relay ladder schematic of Fig. 6-19 can be converted to four rungs of PLC language, as illustrated in Fig. 6-21. In converting these rungs the operation of each rung must be understood. The PB1, PB2, and 1LS references are all programmed using the EXAMINE ON (-| |-) instruction in order to produce the desired logic control action. Also internal relay (address 033) is used to replace control relay ICR. To obtain the desired control logic, all internal relay contacts are programmed using the PLC contact instruction that matches their normal state (NO or NC). Whereas the original hard-wired relay ICR required *four* different contacts, the internal relay uses only *one* contact, which can be examined for an ON or OFF condition as many times as you like. The use of these internal relay equivalents is one of the things that makes the PLC so unique.

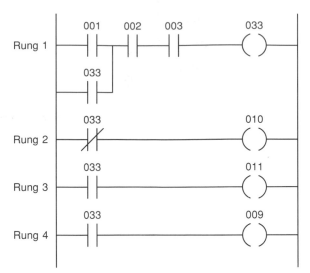

Fig. 6-21 PLC logic ladder program.

REVIEW QUESTIONS

1. (a) Explain the basic operating principle of an electromagnetic control relay.
   (b) Explain the terms *normally open contact* and *normally closed contact* as they apply to this relay.

2. (a) Draw the schematic for an across-the-line ac starter.
   (b) With reference to this schematic, explain the function of each of the following parts:
   (1) Main contact M
   (2) Control contact M
   (3) Starter coil M
   (4) OL relay coils
   (5) OL relay contact
   (c) Explain the difference between the current requirements for the control circuit and power circuit of the starter.

3. (a) Compare the method of operation of each of the following types of switches:
   (1) Manually operated switch
   (2) Mechanically operated switch
   (3) Proximity switch
   (b) What do the abbreviations NO and NC represent when used to describe switch contacts?
   (c) Draw the electrical symbol used to represent each of the following switches:
   (1) NO push button
   (2) NC push button

   (3) Break-make push button
   (4) Single-pole selector switch
   (5) NO limit switch
   (6) NC temperature switch
   (7) NO pressure switch
   (8) NC level switch
   (9) NO proximity switch

4. Draw the electrical symbol used to represent each of the following PLC output control devices:
   (a) Pilot light
   (b) Relay
   (c) Motor starter coil
   (d) OL relay contact
   (e) Alarm
   (f) Heater
   (g) Solenoid
   (h) Solenoid valve
   (i) Motor
   (j) Horn

5. (a) Draw the schematic of a simple electromagnetic latching relay circuit wired to operate a pilot light.
   (b) With reference to this circuit, explain how the pilot light is switched on and off.
   (c) In this circuit, assume that the pilot light was on and power to the circuit lost. When the power is restored will the light be on or off? Why?

6. Explain the difference between a sequential and a combination control process.

PROBLEMS

1. Design and draw the schematic for a conventional hard-wired relay circuit that will perform each of the following circuit functions when an NC push button is pressed:
   - Switch a pilot light on
   - De-energize a solenoid
   - Start a motor running
   - Sound a horn

2. Design and draw the schematic for a conventional hard-wired circuit that will perform the following circuit functions using two break-make push buttons:
   - Turns on light L1 when push button PB1 is pressed.
   - Turns on light L2 when push button PB2 is pressed.
   - Electrically interlock the push buttons so that L1 and L2 cannot both be turned on at the same time.

3. Study the ladder logic program (see Fig. 6-22) and answer the questions that follow:
   (a) Under what condition will the latch rung 1 be TRUE?
   (b) Under what conditions will the unlatch rung 2 be TRUE?
   (c) Under what condition will rung 3 be TRUE?
   (d) When PL1 is on, the relay is in what state (LATCHED or UNLATCHED)?

   (e) When PL2 is on, the relay is in what state (LATCHED or UNLATCHED)?
   (f) If ac power is removed and then restored to the circuit, what pilot light will automatically come on when the power is restored?
   (g) Assume the relay is in its LATCHED state and all three inputs are FALSE. What input change(s) must occur for the relay to switch into its UNLATCHED state?
   (h) If the EXAMINE ON instructions at addresses 001, 002, and 003 are all TRUE, what state will the relay remain in (LATCHED or UNLATCHED)?

4. Design a PLC program and prepare a typical I/O connection diagram and logic ladder program that will correctly execute the hard-wired control circuit (see Fig. 6-23).

5. Design a PLC program and prepare a typical I/O connection diagram and logic ladder program that will correctly execute the hard-wired control circuit (see Fig. 6-24).

6. Design a PLC program and prepare a typical I/O connection diagram and logic ladder program that will correctly execute the hard-wired control circuit (see Fig. 6-25).

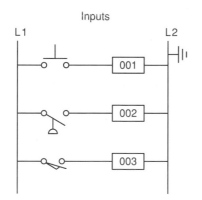
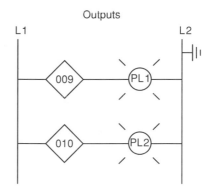
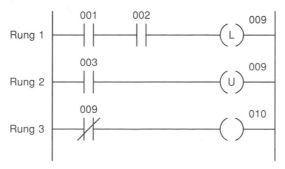

Fig. 6-22

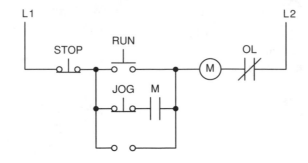

Note: STOP is wired NO.

RUN is wired NO.

JOG is wired using *one* set of
NO contacts.

OL is hard-wired.

Fig. 6-23

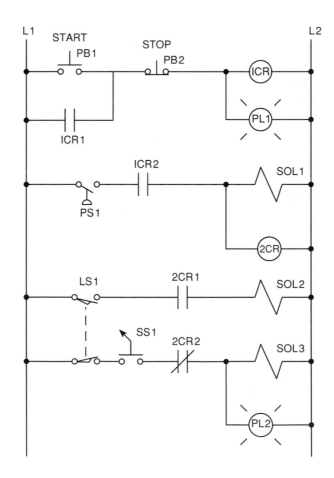

Note: PB1 and PS1 are wired NO.

PB2 is wired NC.

LS1 is wired using *one* set
of NC contacts.

Fig. 6-24

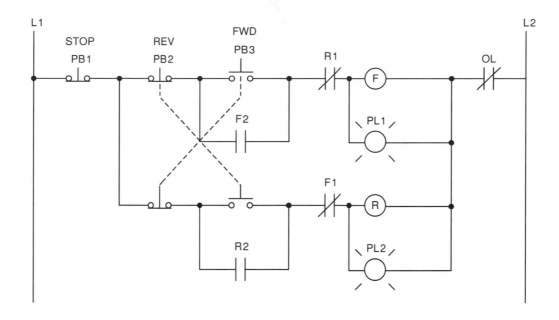

Note: PB1 is wired NC.

PB2 and PB3 are wired using *one* set of NO contacts.

OL is hard-wired

Fig. 6-25

7. Design a PLC program and prepare a typical I/O connection diagram and logic ladder program for the following motor control specifications:
   - A motor must be started and stopped from any one of three START/STOP push button stations.
   - Each START/STOP station contains one NO START button and one NC STOP button.
   - Motor OL contacts are to be hard-wired.

8. Design a PLC program and prepare a typical I/O connection diagram and logic ladder program for the following motor control specifications:

   - Three starters are to be wired so that each starter is operated from its own START/STOP push-button station.
   - A master STOP station is to be included that will trip out all starters when pushed.
   - Overload relay contacts are to be programmed so that an overload on any one of the starters will automatically drop all of the starters.
   - *All* push buttons are to be wired using one set of NO contacts.

7

PROGRAMMING TIMERS

Upon completion of this chapter you will be able to:

- Describe the operation of pneumatic on-delay and off-delay timers
- Describe PLC timer instruction and differentiate between a nonretentive and retentive timer
- Convert fundamental timer relay schematic diagrams to PLC logic ladder programs
- Analyze and interpret typical PLC timer logic ladder programs

7-1 MECHANICAL TIMING RELAY

There are very few industrial control systems that do not need at least one or two timed functions. Mechanical timing relays are used to delay the opening or closing of contacts for circuit control. The operation of a mechanical timing relay is similar to that of a control relay, except that certain of its contacts are designed to operate at a preset time interval, after the coil is energized or de-energized.

Figure 7-1 shows the construction of an on-delay pneumatic (air) timer. The time-delay function depends upon the transfer of air through a restricted orifice. The time-delay period is adjusted by positioning the needle valve to vary the amount of orifice restriction. When the coil is energized, the timed contacts are prevented from opening or closing. However, when the coil is de-energized, the timed contacts return instantaneously to their normal state. This particular pneumatic timer has nontimed contacts in addition to timing contacts. These nontimed contacts are controlled directly by the timer coil, as in a general purpose control relay.

Mechanical timing relays provide time delay through two arrangements. The first arrangement, *on delay* (previously illustrated), provides time delay when the relay is *energized*. The second arrangement, *off delay,* provides time delay when the relay is *de-energized*. Figure 7-2 illustrates the standard relay diagram symbols used for timed contacts.

The circuits of Figs. 7-3, 7-4, 7-5, and 7-6 are designed to illustrate the basic timed-contact functions. In each circuit the time-delay setting of the timing relay is assumed to be 10 s.

7-2 TIMER INSTRUCTIONS

PLC timers are output instructions that provide the same functions as mechanical timing relays. They are used to activate or de-activate a device after a preset interval of time. The timer and counter instructions are the second oldest pair of PLC instructions beside the standard relay instruction discussed in the previous unit. While earlier, first-generation PLC systems did not include these instructions, they are found on all PLCs manufactured today. The number of timers that can be programmed depends upon the model of PLC you are using. However, the availability usually far exceeds the requirement.

There are two methods used to represent a timer within a PLC's logic ladder program. The first depicts the timer instruction as a relay coil similar to that illustrated in Fig. 7-7. The timer is assigned an address as well as being identified as a timer. Also included as part of the timer instruction is the time base of the timer, the timer's preset value or time-delay period, and the accumulated value or current time-delay period for the timer. When the timer rung has logic continuity, the timer begins counting time-based intervals and times until the accumulated

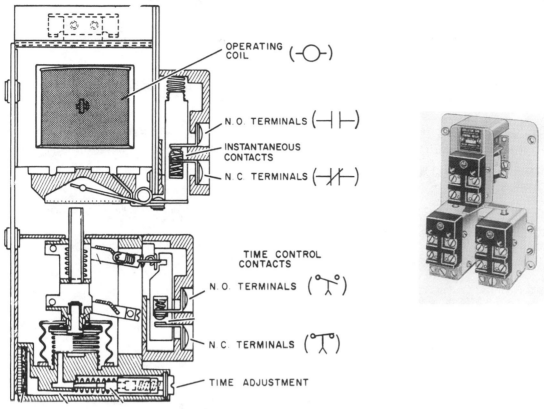

OPERATING COIL (–O–)

N.O. TERMINALS (–| |–)

INSTANTANEOUS CONTACTS

N.C. TERMINALS (–|/|–)

TIME CONTROL CONTACTS

N.O. TERMINALS

N.C. TERMINALS

TIME ADJUSTMENT

Fig. 7-1 Pneumatic ON-DELAY timer. *(Courtesy of Allen-Bradley Company, Inc.)*

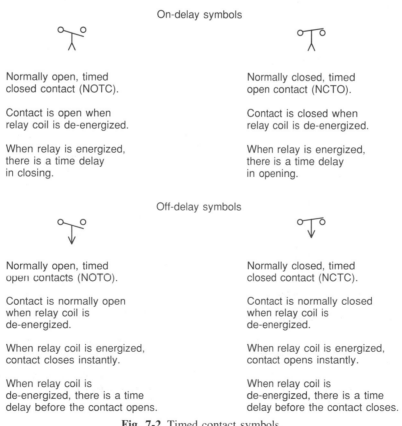

On-delay symbols

Normally open, timed
closed contact (NOTC).

Contact is open when
relay coil is de-energized.

When relay is energized,
there is a time delay
in closing.

Normally closed, timed
open contact (NCTO).

Contact is closed when
relay coil is de-energized.

When relay is energized,
there is a time delay
in opening.

Off-delay symbols

Normally open, timed
open contacts (NOTO).

Contact is normally open
when relay coil is
de-energized.

When relay coil is energized,
contact closes instantly.

When relay coil is
de-energized, there is a time
delay before the contact opens.

Normally closed, timed
closed contact (NCTC).

Contact is normally closed
when relay coil is
de-energized.

When relay coil is energized,
contact opens instantly.

When relay coil is
de-energized, there is a time
delay before the contact closes.

Fig. 7-2 Timed contact symbols.

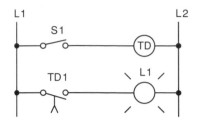

Sequence of operation:

S1 open, TD de-energized, TD1 open, L1 is off.

S1 closes, TD energizes, timing period starts, TD1 still open, L1 still off.

After 10 s, TD1 closes, L1 is switched on.

S1 is opened, TD de-energizes, TD1 opens instantly, L1 is switched off.

Fig. 7-3 ON-DELAY timer circuit (NOTC contact).

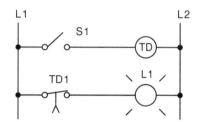

Sequence of operation:

S1 open, TD de-energized, TD1 closed, L1 is on.

S1 closes, TD energizes, timing period starts, TD1 still closed, L1 still on.

After 10 s, TD1 opens, L1 is switched off.

S1 is opened, TD de-energizes, TD1 closes instantly, L1 is switched on.

Fig. 7-4 ON-DELAY timer circuit (NCTO contact).

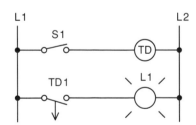

Sequence of operation:

S1 open, TD de-energized, TD1 open, L1 is off.

S1 closes, TD energizes, TD1 closes instantly, L1 is switched on.

S1 is opened, TD de-energizes, timing period starts, TD1 still closed, L1 still on.

After 10 s, TD1 opens, L1 is switched off.

Fig. 7-5 OFF-DELAY timer circuit (NOTO contact).

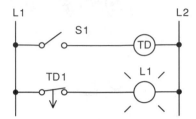

Sequence of operation:

S1 open, TD de-energized, TD1 closed, L1 is on.

S1 closes, TD energizes, TD1 opens instantly, L1 is switched off.

S1 is opened, TD de-energizes, timing period starts, TD1 still open, L1 still off.

After 10 s, TD1 closes, L1 is switched on.

Fig. 7-6 OFF-DELAY timer circuit (NCTC contact).

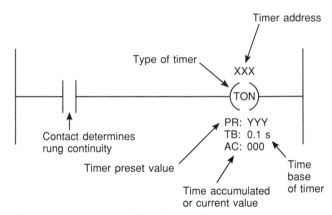

Fig. 7-7 Coil-formatted timer instruction.

value equals the preset value. When the accumulated time equals the preset time, the output is energized and the timed output contact associated with the output is closed. The timed contact can be used throughout the program as an NO or NC contact as many times as you wish.

The second timer format is referred to as a *block format*. Figure 7-8 illustrates a generic block format. The timer block has two input conditions associated with it, namely the *control* and *reset*. The control line controls the actual timing operation of the timer. Whenever this line is TRUE or power is supplied to this input, the timer will time. Removal of power from the control line input halts the further timing of the timer.

The reset line resets the timer's accumulated value to zero. Some manufacturers require that *both* the control and reset lines be TRUE for the timer to time; removal of power from the reset input resets the timer to zero. Other manufacturers' PLCs require power flow for the control input *only* and no power flow on the reset input for the

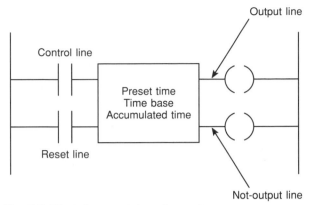

Fig. 7-8 Block-formatted timer instruction.

timer to operate. For this type of timer operation the timer is reset whenever the reset input is TRUE.

The timer instruction block contains information pertaining to the operation of the timer. This includes the preset time, the time base of the timer, and the current or accumulated time.

All block-formatted timers provide at least one output signal from the timer. When a single output is provided, it is used to signal the completion of the timing cycle. For dual-output timer instructions, the second output signal operates in the reverse mode. Whenever the timer has *not* reached its timed-out state, the second output is on while the first output remains off. As soon as the timer reaches its timed-out state, the second output is turned off and the first output is turned on.

7-3 ON-DELAY TIMER INSTRUCTION

The ON-DELAY timer instruction emulates the operation of the pneumatic electromechanical ON-DELAY timer. Figure 7-9 illustrates the generic programming of an ON-DELAY timer that uses the block-formatted instruction. This timer is activated by closing the switch connected to input 001. The preset time for this timer is 10 s, at which time output 009 will be energized. When the switch is closed, the timer begins counting, and counts until the accumulated time equals the preset value; the output is then energized. If the switch is opened before the timer is timed out, the accumulated time is automatically reset to zero. This timer configuration is termed *nonretentive* since loss of power flow to the timer causes the timer instruction to reset. This timing operation is that of an ON-DELAY timer, since the lamp is switched on 10 s after the switch has been actuated from the OFF to the ON position.

Timers may or may not have an instantaneous output signal associated with them. If an instantaneous output signal is required from a timer and it is not provided as part of the timer instruction, an equivalent instantaneous contact instruction can be programmed using an internally referenced relay coil.

Figure 7-10 shows an application of this technique. According to the relay ladder schematic diagram, coil M is to be energized 5 s after the START push button is pressed. Contact TD1 is the instantaneous contact, and contact TD2 is the timed contact. The logic ladder pro-

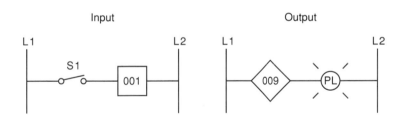

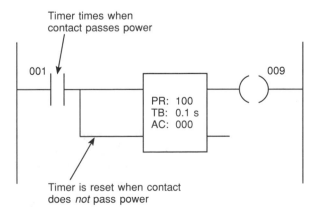

Fig. 7-9 ON-DELAY programmed timer.

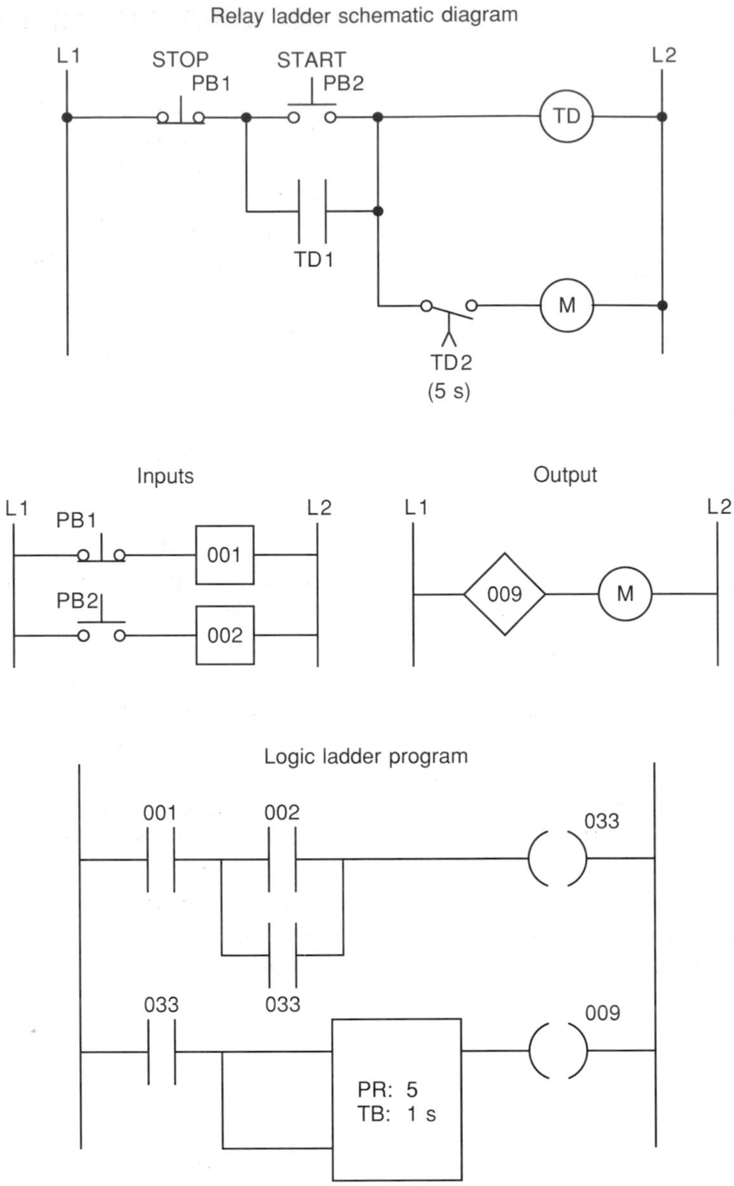

Relay ladder schematic diagram

Inputs

Output

Logic ladder program

Fig. 7-10 ON-DELAY timer with instantaneous output programming.

gram shows that a contact instruction referenced to an internal relay is now used to operate the timer. The instantaneous contact is referenced to the internal relay coil, while the time-delay contact is referenced to the timer output coil.

Figure 7-11 shows an application for an on-delay timer that uses an NCTO contact. This circuit is used as a warning signal when moving equipment, such as a conveyor motor, is about to be started. According to the relay ladder schematic diagram, coil CR is energized when the START push button PB1 is momentarily actuated. As a result, contact CR1 closes to seal in CR, contact CR2 closes to energize timer coil TD, and contact CR3 closes to sound the horn. After a 10-s time-delay period, timer contact TD1 opens to automatically switch the horn off. The logic ladder program shows how the circuit could be programmed using a PLC with a coil-formatted timer in-

struction.

Timers are often used as part of automatic sequential control systems. Figure 7-12 shows how a series of motors can be started automatically with only one START/STOP control station. According to the relay ladder schematic diagram, lube-oil pump motor starter coil 1M is energized when the START push button PB2 is momentarily actuated. As a result, 1M1 control contact closes to seal in 1M, and the lube-oil pump motor starts. When the lube-oil pump builds up sufficient oil pressure, the lube-oil pressure switch PS1 closes. This in turn energizes coil 2M to start the main drive motor, and energizes coil TD to begin the time-delay period. After the preset time-delay period of 15 s, TD1 contact closes to energize coil 3M and start the feed motor. The logic ladder program shows how the circuit could be programmed using a PLC.

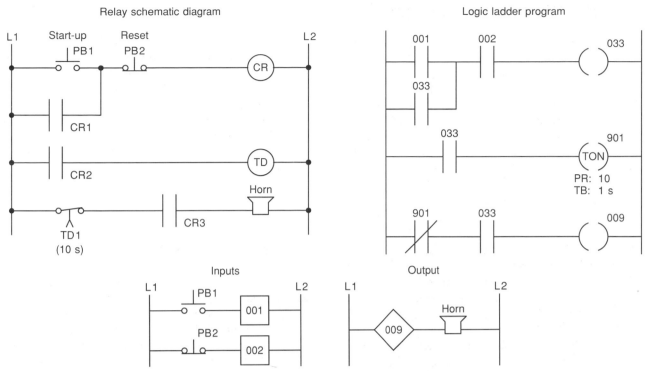

Fig. 7-11 Starting-up warning signal circuit.

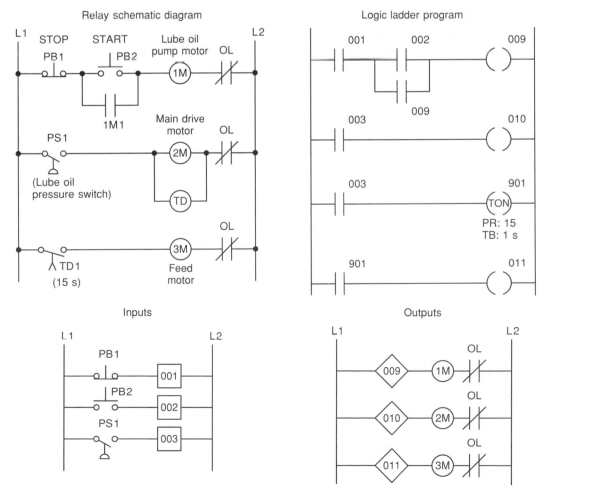

Fig. 7-12 Automatic sequential control system.

7-4 OFF-DELAY TIMER INSTRUCTION

The OFF-DELAY timer instruction emulates the operation of the pneumatic electromechanical OFF-DELAY timer. Figure 7-13 illustrates the generic programming of an OFF-DELAY timer that uses the coil-formatted timer instruction. If logic continuity is *lost,* the timer begins counting time-based intervals until the accumulated time equals the programmed preset value. When the switch connected to input 001 is first closed, timed output 009 is set to 1 immediately and so the lamp is switched on. If this switch is now opened, logic continuity is lost and the timer begins counting. After 10 s, when the accumulated time equals the preset time, the output is reset to 0 and the lamp switches off. If logic continuity is gained before the timer is timed out, the accumulated time is reset to zero. For this reason this timer is also classified as being nonretentive.

Figure 7-14 shows how a relay circuit with a pneumatic off-delay timer could be programmed using a PLC. According to the relay ladder schematic diagram, when power is first applied (limit switch LS1 open), motor starter coil 1M is energized and the green pilot light is on. At the same time, motor starter coil 2M is de-energized, and the red pilot light is OFF.

When limit switch LS1 closes, off-delay timer coil TD energizes. As a result, timed contact TD1 opens to de-energize motor starter coil 1M, timed contact TD2 closes to energize motor starter coil 2M, instantaneous contact TD3 opens to switch green light off, and instantaneous contact TD4 closes to switch red light on. The circuit remains in this state as long as limit switch LS1 is closed.

When limit switch LS1 is opened, the off-delay timer coil TD de-energizes. As a result, the time-delay period is started, instantaneous contact TD3 closes to switch the green light on, and instantaneous contact TD4 opens to switch the red light off. After a 5-s time-delay period, timed contact TD1 closes to energize motor starter 1M, and timed contact TD2 opens to de-energize motor starter 2M. An internal relay and timer instruction are used to implement the program.

7-5 RETENTIVE TIMER

A *retentive timer* is one that accumulates time whenever the device receives power, and maintains the current time should power be removed from the device. Once the device accumulates time equal to its preset value, the contacts of the device change state. Loss of power to the device after reaching its preset value does not affect the state of the contacts. The retentive timer must be *intentionally reset* with a separate signal in order for the accumulated time to be reset and for the contacts of the device to return to their shelf state.

Figure 7-15 illustrates the action of a motor-driven electromechanical retentive timer used in some appliances. The shaft-mounted cam is driven by a motor. Once power is applied, the motor starts turning the shaft and cam.

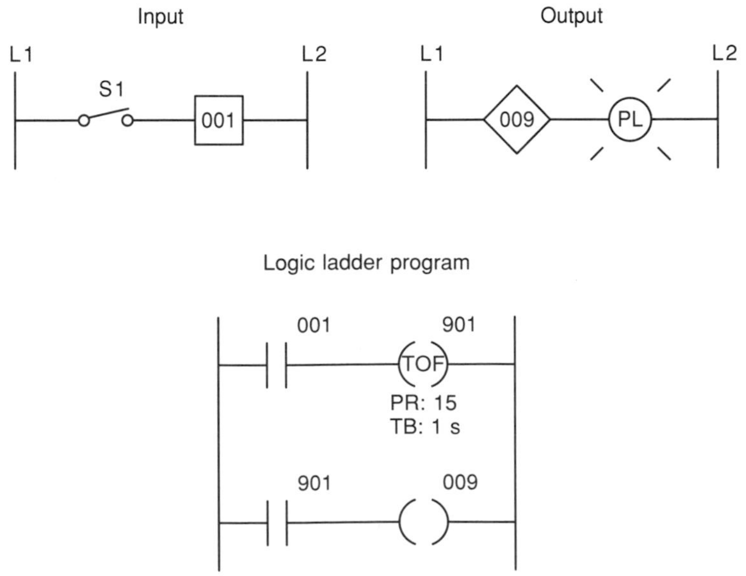

Fig. 7-13 OFF-DELAY programmed timer.

Relay schematic diagram

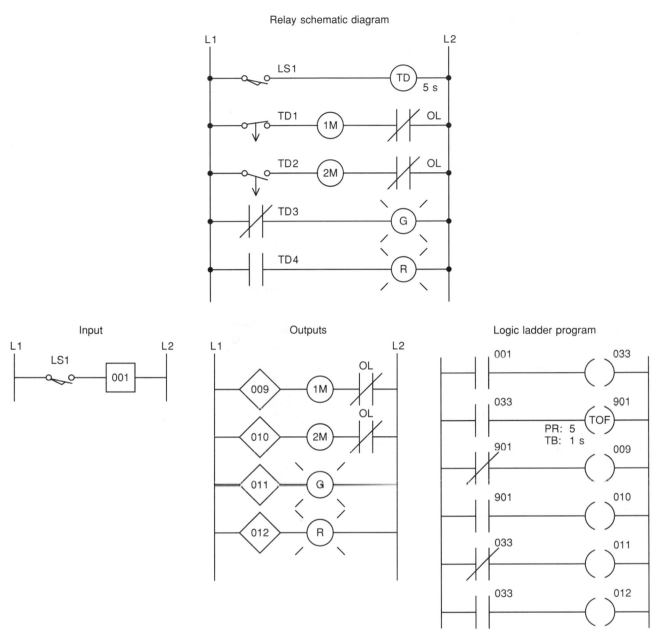

Fig. 7-14 Programming a pneumatic OFF-DELAY timer circuit.

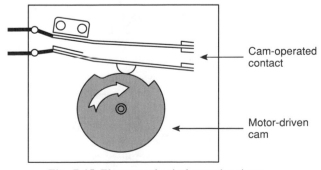

Fig. 7-15 Electromechanical retentive timer.

The positioning of the lobes of the cam and the gear reduction of the motor determine the time it takes for the motor to turn the cam far enough to activate the contacts. If power is removed from the motor, the shaft stops but *does not reset*.

The PLC-programmed retentive ON-DELAY timer (RTO) operates in the same way as the nonretentive ON-DELAY timer (TON) with one major exception. This is a RETENTIVE TIMER RESET (RTR) instruction. Unlike the TON, the RTO will hold its accumulated value when the timer rung goes FALSE and will continue timing where it left off when the timer rung goes TRUE again. This timer must be accompanied by a timer RESET instruction to reset the accumulated value of the timer to zero. The RTR instruction is the *only* automatic means of resetting the accumulated value of a retentive timer. The RTR instruction must be addressed to the same word as the RTO instruction. Normally, if any RTR rung path has logic continuity, then the accumulated value of the referenced timer is reset to zero.

Figure 7-16 shows a typical PLC program for an RTO along with a timing chart for the circuit. The timer will start to time when the NO push button PB1 referenced to input 11306 is closed. If the push button is opened after 3 s, the timer accumulated value stays at 003. When the push button is closed again, the timer picks up the time at 3 s and continues timing. When the accumulated value equals the preset value 009, bit 15 of word 052 is set to 1 and output 01004 and the light are both on.

Since the retentive timer does not reset to zero when the timer is de-energized, reset rung 3 must be used to reset the timer. This rung consists of an NO push button PB2 referenced to input 11307 and an RTR instruction. The RTR instruction is given the same address (052) as the RTO. When push button PB2 closes, RTR resets the accumulated time to zero and changes timed bit 15 of word 052 to 0, turning that light off.

The program drawn in Fig. 7-17 illustrates a practical application for an RTO. The operation of the RTO timer is to detect whenever a piping system has sustained a *cumulative* overpressure condition of 60 s. At that point, a horn is automatically sounded to call attention to the malfunction. When they are alerted, maintenance personnel can silence the alarm by switching the key switch S1 to the RESET (contact closed) position. After the problem has been corrected, the alarm system can be reactivated by switching the key switch to the ON (contact open) position.

A retentive OFF-DELAY timer is programmed in the same manner as an RTO. Both maintain their accumulated time value even if logic continuity is lost before the timer is timed out or if power is lost. These retentive timers do *not* have to be completely timed out in order to be reset. Rather, such a timer can be reset at any time during its operation. It should be noted that the reset input to the timer will override the control input of the timer, to reset it even though the control input to the timer has logic continuity.

7-6 CASCADING TIMERS

The programming of two or more timers together is called *cascading*. Timers can be interconnected, or cascaded, to satisfy any required control logic. Figure 7-18 shows how three motors can be automatically started in sequence with a 20-s time delay between each motor start-up. According to the relay ladder schematic diagram, motor starter coil 1M is energized when the START push button PB2 is momentarily actuated. As a result, motor 1 starts, contact 1M1 closes to seal in 1M, and timer coil 1TD is energized to begin the first time-delay period. After the preset time period of 20 s, 1TD1 contact closes to energize motor starter coil 2M. As a result, motor 2 starts and timer coil 2TD is energized to begin the second time-delay period. After the preset time period of 20 s, 2TD1 contact closes to energize motor starter coil 3M, and so motor 3 starts. The logic ladder program shows how the circuit could be programmed using a PLC. Note that two internal timers are used and the output of the first timer is used to control the input logic to the second timer.

Two timers can be interconnected to form an oscillator circuit. The oscillator logic is basically a timing circuit programmed to generate periodic output pulses of any duration. Figure 7-19 shows the program for an annunciator flasher circuit. Two internal timers form the oscillator circuit, which generates a timed, pulsed output. The oscillator circuit output is programmed in series with the alarm condition. If the alarm condition (temperature pressure, or limit switch) is TRUE, the appropriate output indicating light will flash. Note that any number of alarm conditions could be programmed using the same flasher circuit.

At times you may require a time-delay period longer than the maximum preset time allowed for the single timer instruction of the PLC being used. When this is the case, the problem can be solved by simply cascading timers as illustrated in Fig. 7-20. The type of timer programmed for this example is a TON. The total time delay required is 1200 s (999 + 201 s, or 20 min). The first timer is programmed for its maximum preset time of 999 s and begins timing when field switch S1 is closed. When it completes its time-delay period, 999 s later, internally referenced contacts 901 will close. This action in turn will activate the second timer, which is preset for the remaining 201 s of the total 1200-s time delay. Once the second timer reaches its preset time, internally referenced contact 902 closes to turn the light on and indicate the completion of the full 20-min time delay. Opening of field switch S1 at any time will reset both timers and switch the light off.

Inputs

Output

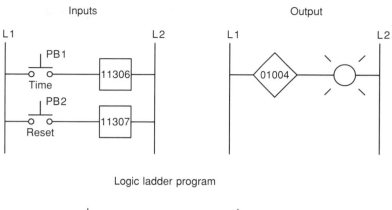

Logic ladder program

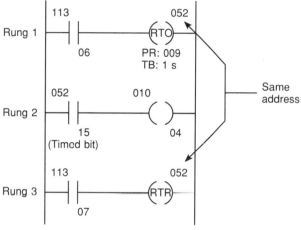

Timing chart

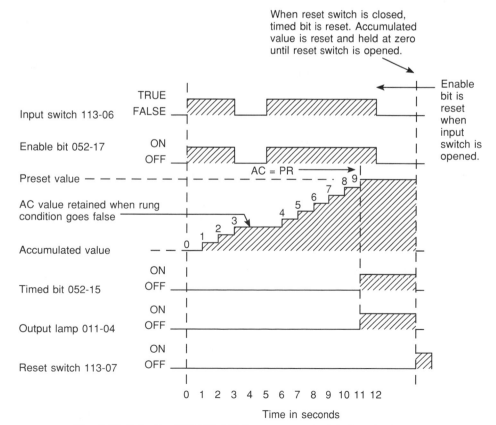

Fig. 7-16 Retentive ON-DELAY timer program and timing chart.

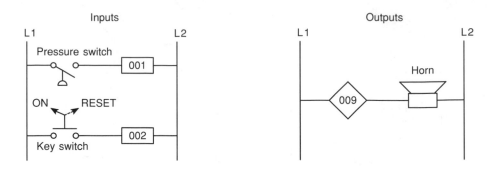

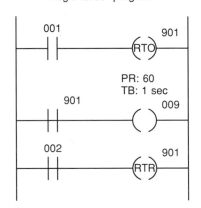

Logic ladder program

Fig. 7-17 Retentive ON-DELAY alarm program.

Relay schematic diagram

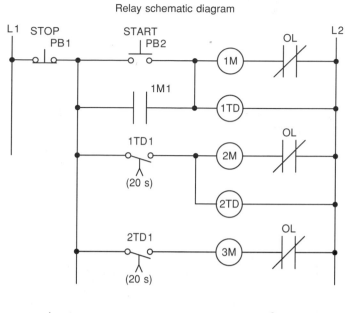

Inputs

Outputs

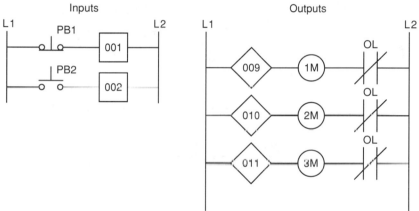

Logic ladder program

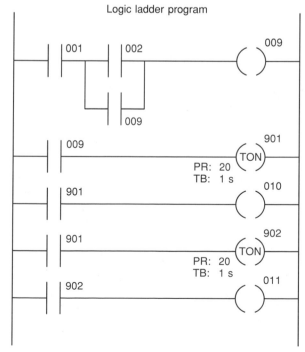

Fig. 7-18 Sequential time-delayed motor-starting circuit.

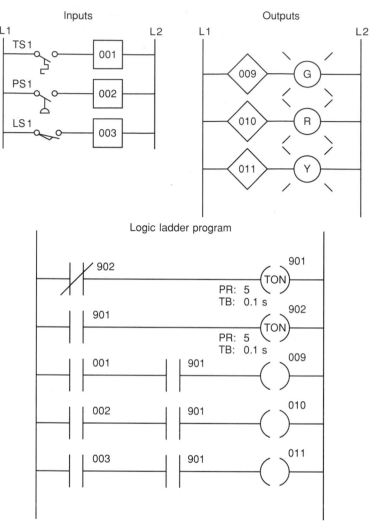

Fig. 7-19 Annunciator flasher circuit.

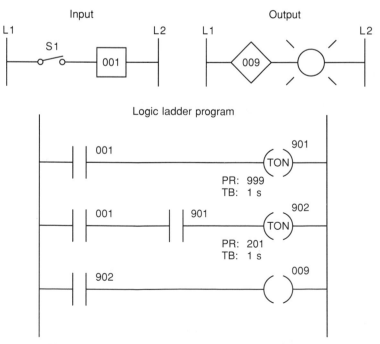

Fig. 7-20 Cascading of timers for longer time delays.

Programmable controller with annunciator I/O system. *(Courtesy of Square D Company)*

REVIEW QUESTIONS

1. Explain the difference between the timed and instantaneous contacts of a pneumatic timer.

2. Draw the symbol and explain the operation of each of the following timed contacts of a pneumatic timer:
   (a) On-delay timer—NOTC contact
   (b) On-delay timer—NCTO contact
   (c) Off-delay timer—NOTO contact
   (d) Off-delay timer—NCTC contact

3. State five pieces of information usually associated with a PLC timer instruction.

4. When is the output of a programmed timer energized?

5. What are the two methods commonly used to represent a timer within a PLC's logic ladder program?

6. (a) Explain the difference between the operation of a nonretentive and a retentive timer.
   (b) Explain how the accumulated count of programmed retentive and nonretentive timers is reset to zero.

PROBLEMS

1. (a) With reference to the relay schematic diagram (Fig. 7-21), state the status of each light (ON or OFF) after each of the following sequential events:
   (1) Power is first applied and switch S1 is open.
   (2) Switch S1 has just closed.
   (3) Switch S1 has been closed for 5 s.
   (4) Switch S1 has just opened.
   (5) Switch S1 has been opened for 5 s.
   (b) Design a PLC program and prepare a typical I/O connection diagram and logic ladder program that will correctly execute this hard-wired control circuit.

2. Design a PLC program and prepare a typical I/O connection diagram and logic ladder program (Fig. 7-22) that will correctly execute a hard-wired control circuit.

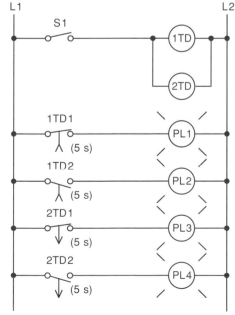

Relay schematic diagram

Fig. 7-21

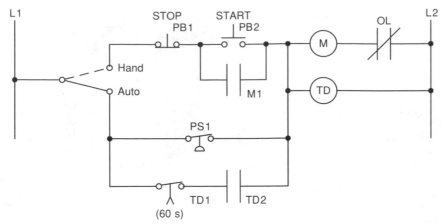

Fig. 7-22

3. Study the logic ladder program and answer the questions that follow (Fig. 7-23).
   (a) What type of timer has been programmed?
   (b) What is the length of the time-delay period?
   (c) What is the value of the accumulated time when power is first applied?
   (d) Which address is an internal relay instruction?
   (e) When does the timer start timing?
   (f) When does the timer stop timing and reset itself?
   (g) When input 001 is first turned on, which rungs are TRUE and which rungs are FALSE?
   (h) When input 001 is first turned on, state the status (ON or OFF) of each output.
   (i) When the timer's accumulated value equals the preset value, which rungs are TRUE and which rungs are FALSE?
   (j) When the timer's accumulated value equals the preset value, state the status (ON or OFF) of each output.
   (k) Suppose that rung 1 is true for 5 s and then power is lost. What will the accumulated value of the counter be when power is restored?

4. Study the logic ladder program (Fig. 7-24) and answer the questions that follow.
   (a) What type of timer has been programmed?
   (b) What is the length of the time-delay period?
   (c) When does the timer start timing?
   (d) When is the timer reset?
   (e) When will rung 3 be TRUE?
   (f) When will rung 5 be TRUE?
   (g) When will output 012 be energized?
   (h) Assume your accumulated time value is up to 020 and power to your system is lost. What will your accumulated time value be when power is restored?
   (i) What happens if inputs 001 and 002 are both TRUE at the same time?

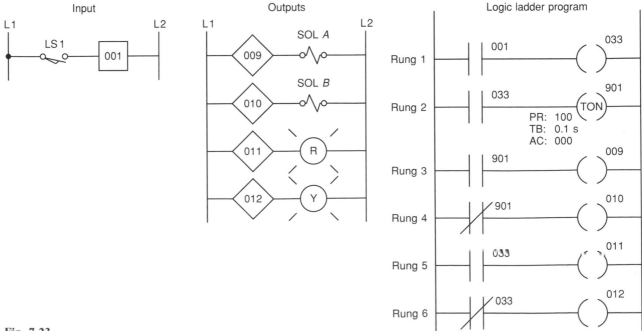

Fig. 7-23

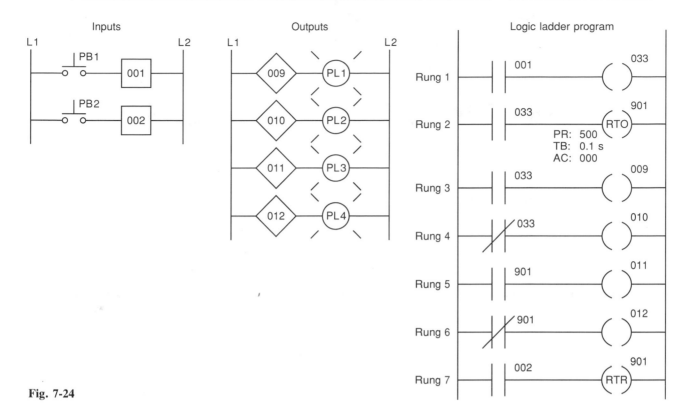

Fig. 7-24

5. Study the logic ladder program (Fig. 7-25) and answer the questions that follow.
   (a) What is the purpose of interconnecting the two timers?
   (b) How much time must elapse before output 009 is energized?
   (c) What two conditions must be satisfied in order for timer 902 to start timing?
   (d) Assume that output 009 is on and power to the system is lost. When power is restored, what will the status of this output be?
   (e) When input 002 is on, what will happen?
   (f) When input 001 is on, how much accumulated time must elapse before rung 2 will be TRUE?
   (g) Assume that the two timers are to be programmed as nonretentive timers. What changes would have to be made in the logic ladder program?

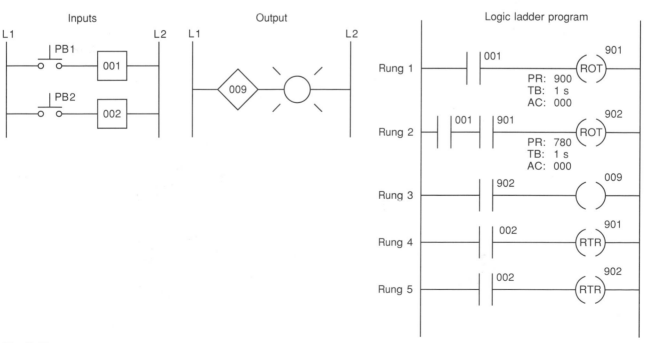

Fig. 7-25

8

PROGRAMMING COUNTERS

Upon completion of this chapter you will be able to:

- List and describe the functions of PLC counter instructions
- Describe the operating principle of a transitional, or one-shot, contact
- Analyze and interpret typical PLC counter logic ladder programs

8-1 COUNTER INSTRUCTIONS

Programmed counters can serve the same function as mechanical counters. Figure 8-1 shows the construction of a simple mechanical counter. Every time the actuating lever is moved over, the counter adds one number, while the actuating lever returns automatically to its original position. Resetting to zero is by a push button located on the side of the unit.

Programmed counters can count up, count down, or be combined to count up and down. While the majority of counters used in industry are up-counters, numerous applications require the implementation of down-counters or of combination up/down-counters. Practically every

PLC model offers some form of counter instruction as part of its instruction set.

Counters are similar to timers, except that they do not operate on an internal clock, but are dependent upon external or program sources for counting. As was the case with the timer instruction, there are two methods used to represent a counter within a PLC's logic ladder program. Similar to the timer, a coil programming format, similar to that illustrated in Fig. 8-2, is used by many manufacturers. The counter is assigned an address as well as being identified as a counter. Also included as part of the counter instruction is the counter's *preset value* as well as the current *accumulated count* for the counter. The up-counter increments its accumulated value by 1 each

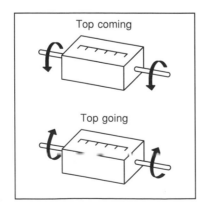
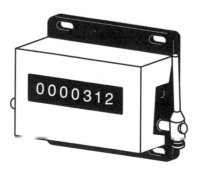

Fig. 8-1 Mechanical counter.

96

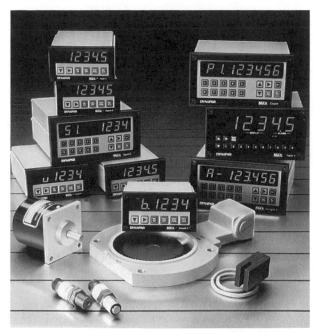

Programmed counter. *(Courtesy of Dynapar Corporation, Gurnee, Illinois.)*

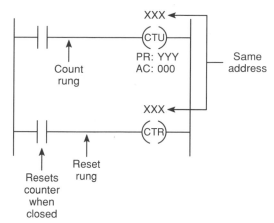

Fig. 8-3 Coil-formatted counter and RESET instructions.

time the counter rung makes a FALSE-to-TRUE transition. When the accumulated count equals the preset count, the output is energized, and the counter output is closed. The counter contact can be used throughout the program as an NO or NC contact as many times as you wish.

A COUNTER RESET instruction, which permits the counter to be reset, is also used in conjunction with the counter instruction. Up-counters are always reset to zero. Down-counters may be reset to zero or to some preset value. Some manufacturers include the reset function as a part of the general counter instruction, while others dedicate a separate instruction for resetting of the counter. Figure 8-3 shows a generic coil-formatted counter instruction with a separate instruction for resetting of the counter. When programmed, the counter reset coil (CTR) is given the *same* reference address as the counter (CTU)

that it is to reset. The reset instruction is activated whenever the CTR rung condition is TRUE.

The second counter format is referred to as a *block format*. Figure 8-4 illustrates a generic block-formatted counter. The instruction block indicates the type of counter (up or down) along with the counter's preset value and accumulated or current value. The counter has two input conditions associated with it; namely the COUNT and RESET. All PLC counters operate, or count, on the leading edge of the input signal. The counter will either increment or decrement whenever the count input transfers from an OFF state to an ON state. The counter *will not* operate on the trailing edge, or ON-to-OFF transition, of the input condition.

Some manufacturers require the reset rung or line to be TRUE to reset the counter, while others require it to be FALSE to reset the counter. For this reason it is wise to consult the PLC's operator's manual before attempting any programming of counter circuits.

Most PLC counters are normally retentive. That is to say, whatever count was contained in the counter at the time of a processor shutdown will be restored to the counter on power-up. The counter may be reset, however, if the reset condition is activated at the time of power restoration.

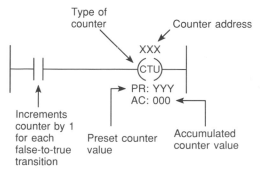

Fig. 8-2 Coil-formatted counter instruction.

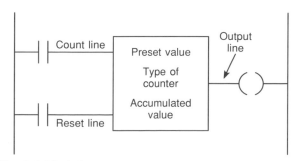

Fig. 8-4 Block-formatted counter instruction.

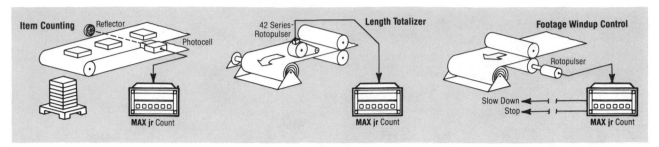

Counter applications. *(Courtesy of Dynapar Corporation, Gurnee, Illinois)*

8-2 UP-COUNTER

The up-counter output instruction will increment by 1 each time the counted event occurs. Figure 8-5 shows the program and timing diagram for a simple up-counter. This control application is designed to turn the red pilot light on and the green pilot light off after an accumulated count of 7. Operating push button PB1 provides the OFF-to-ON transition pulses that are counted by the counter. The preset value of the counter is set for 007. Each FALSE-to-TRUE transition of rung 1 increases the counter's accumulated value by 1. After 7 pulses, or counts, when the preset counter value equals the accumulated counter value, output 901 is energized. As a result, rung 2 becomes TRUE, energizing output 009 to switch the red pilot light on. At the same time, rung 3 becomes FALSE, de-energizing output 010 to switch the green pilot light off. The counter is reset by closing push

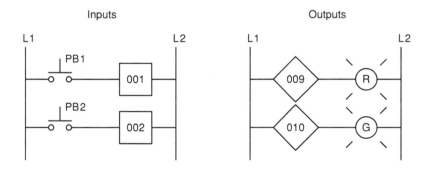

Inputs   Outputs

Logic ladder program

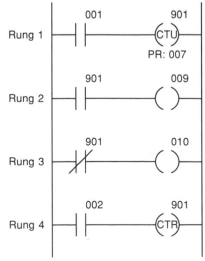

Fig. 8-5 Simple up-counter program.

Timing diagram

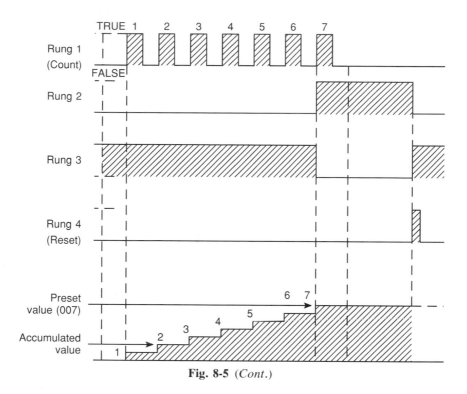

Fig. 8-5 (*Cont.*)

button PB2, which makes rung 4 TRUE and resets the accumulated count to zero. Counting can resume when rung 4 goes FALSE again.

Figure 8-6 shows the program for a *one-shot*, or *transitional, contact circuit* that is often used to automatically clear or reset a counter. The program is designed to generate an output pulse that, when triggered, goes on for the duration of one program scan and then goes off. The one-shot can be triggered from a momentary signal, or one that comes on and stays on for some time. Whichever signal is used, the one-shot is triggered by the leading edge (OFF-to-ON) transition of the input signal. It stays on for one scan and goes off. It stays off until the trigger goes off, and then comes on again. The one-shot is perfect for resetting both counters and timers since it stays on for one scan only.

Some PLCs provide transitional contacts or one-shot instruction in addition to the standard NO and NC contact instructions. The transitional contact (Fig. 8-7a) is programmed to provide a one-shot pulse when the referenced trigger signal makes a positive (OFF-to-ON) transition. This contact will close for exactly one program scan whenever the trigger signal goes from OFF to ON. The contact will allow logic continuity for one scan and then open, even though the triggering signal may stay on. The ON-to-OFF transitional contact (Fig. 8-7b) pro-

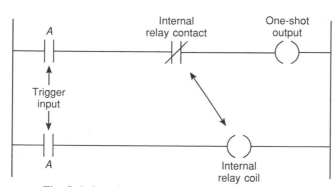

Fig. 8-6 One-shot, or transitional, contact program.

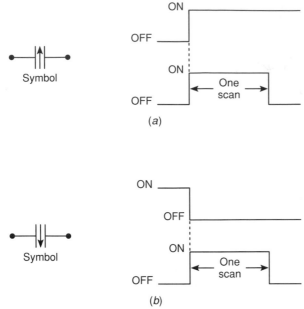

Fig. 8-7 The two types of transitional contact. (*a*) OFF-to-ON transitional contact. (*b*) ON-to-OFF transitional contact.

vides the same operation as the OFF-to-ON transitional contact instruction, except that it allows logic continuity for a single scan whenever the trigger signal goes from an ON to an OFF state.

The conveyor motor PLC program of Fig. 8-8 illustrates the application of an up-counter along with a programmed one-shot reset circuit. The counter counts the number of cases coming off the conveyor. When the total number of cases reaches 50, the conveyor motor stops

automatically. The trucks being loaded will only take a total of 50 cases of this particular product; however, the count can be changed for different product lines. A proximity switch is used to sense the passage of cases.

The sequential task is as follows:

1. START button is pressed to start the conveyor motor.
2. Cases move past proximity switch and increment the counter's accumulated value.
3. After a count of 50 the conveyor motor stops automatically and the counter's accumulated value is reset to zero.
4. The conveyor motor can be stopped and started manually at any time without loss of the accumulated count.
5. The accumulated count of the counter can be reset manually at any time by means of the COUNT RESET button.

The alarm monitor PLC program of Fig. 8-9 illustrates the application of an up-counter used in conjunction with the timed oscillator circuit studied in the previous unit. The operation of the alarm monitor is as follows:

1. The alarm is triggered by the closing of liquid level switch LS1.
2. The light will flash whenever the alarm condition is triggered and has not been acknowledged even if the alarm condition clears in the mean time.
3. The alarm is acknowledged by closing selector switch SS1.
4. The light will operate in the steady ON mode when the alarm trigger condition still exists but has been acknowledged.

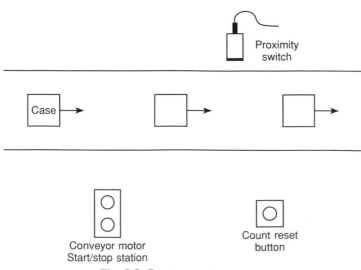

Process flow diagram

Fig. 8-8 Conveyor motor program.

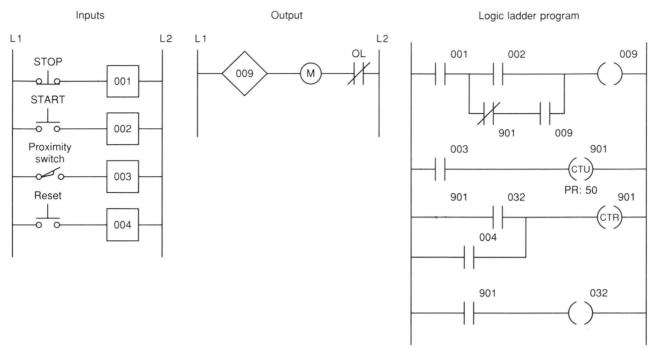

Fig. 8-8 *(Cont.)*

8-3 DOWN-COUNTER

The down-counter output instruction will count down or decrement by 1 each time the counted event occurs. Each time the down-count event occurs, the accumulated value is decremented. Normally the down-counter is used in conjunction with the up-counter to form an up/down-counter. Figure 8-10 shows the program and timing diagram for a simple, block-formatted up/down-counter. Separate count-up and count-down inputs are provided. Assuming the preset value of the counter is 3 and the accumulated count is zero, pulsing the count-

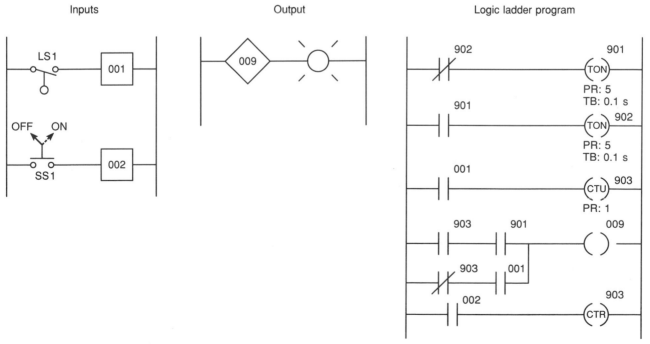

Fig. 8-9 Alarm monitor.

Inputs

Output

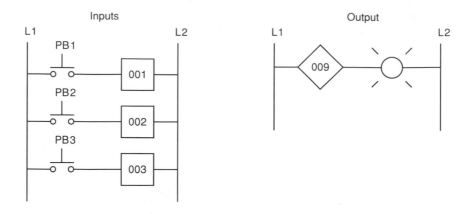

Logic ladder program

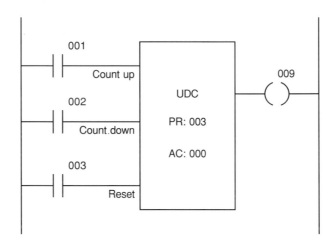

Timing diagram

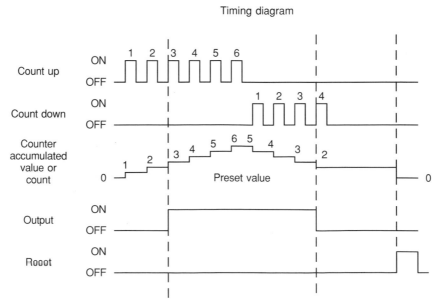

Fig. 8-10 Simple up/down-counter program.

up input (PB1) three times will switch the output light from OFF to ON. This particular PLC counter keeps track of the number of counts received above the preset value. As a result, three additional pulses of the count-up input (PB1) produce an accumulated value of 6 but no change in the output. If the count-down input (PB2) is now pulsed four times, the accumulated count is reduced to 2 (6 − 4). As a result, the accumulated count drops below the preset count and the output light switches from ON to OFF. Pulsing the reset input (PB3) at any time will reset the accumulated count to zero and turn the output light off.

Not all counter instructions count in the same manner. Some up-counters count only to their preset values and additional counts are ignored. Other up-counters keep track of the number of counts received above the counter's preset value. Conversely some down-counters will simply count down to zero and no further. Other down-counters may count below zero and begin counting down from the largest preset value that can be set for the PLC's counter instruction. For example, a PLC up/down-counter that has a maximum counter preset limit of 999 may count up as follows: 997, 998, 999, 000, 001, 002, etc. The same counter would count down in the following manner: 002, 001, 000, 999, 998, 997, etc.

A typical application for an up/down-counter could be to keep count of the cars that enter and leave a parking garage. As a car enters, it triggers the up-counter output instruction and increments the accumulated count by 1. Conversely, as a car leaves, it triggers the down-counter output instruction and decrements the accumulated count by 1. Since both the up- and down-counters have the

same address, the accumulated value will be the same in both. Whenever the accumulated value equals the preset value, the counter output is energized to light up the LOT FULL sign. Figure 8-11 shows a typical PLC program that could be used to implement the circuit. A coil-formatted type of counter instruction is used, and a RESET button has been provided to reset the accumulated count.

8-4 CASCADING COUNTERS

The maximum count in most controllers is 9999 per counter instruction. Depending on the application, it may be necessary to count events that exceed the maximum number allowable per counter instruction. One way of accomplishing this is by interconnecting, or cascading, two counters. The program of Fig. 8-12 illustrates the application of the technique. In this program the output of the first counter is programmed into the input of the second counter. The status bits of both counters are programmed in series to produce an output. These two counters allow twice as many counts to be measured.

Some control systems incorporate a 24-h clock to display the time of day or for the logging of data pertaining to the operation of the process. The logic used to implement a clock as part of a PLC's program is straightforward and simple to accomplish. A single timer and counter instructions is all you need.

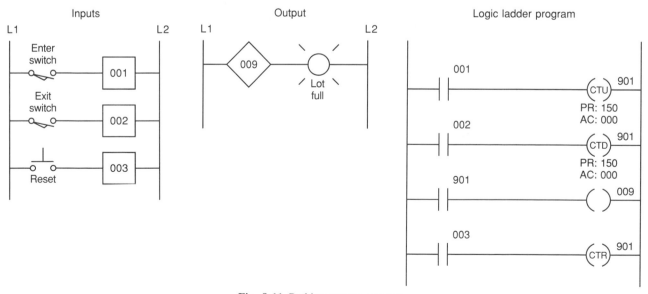

Fig. 8-11 Parking garage counter.

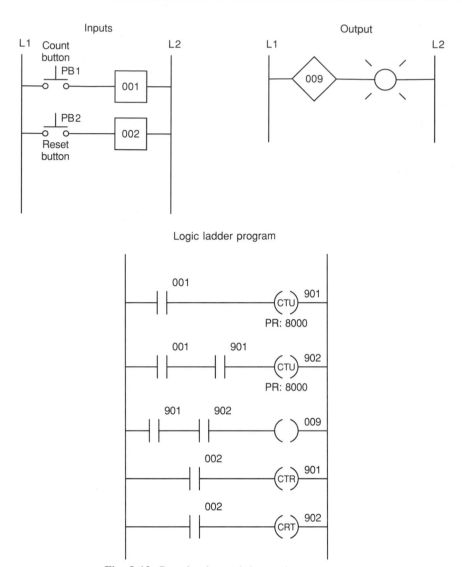

Fig. 8-12 Counting beyond the maximum count.

Figure 8-13 illustrates a timer-counter program that produces a time-of-day clock measuring time in hours and minutes. A timer instruction is programmed first with a preset value of 60 s. This timer times for a 60-s period, after which internal coil 901 is activated. The energization of coil 901 causes the counter of rung 2 to increment one count. On the next processor scan the timer is reset and begins timing again. The counter of rung 2 is preset to 60 counts, and each time the timer completes its time delay, its count is incremented. When this counter reaches its preset value of 60, internal coil 902 is energized. Energization of coil 902 increments the second counter programmed in rung 3. The counter of rung 3 is preset for 24 counts. Whenever coil 902 is activated, it also resets the first counter to begin the 60-count sequence again. Whenever the second counter reaches its preset value of 24, internal coil 903 is energized to reset itself. The time of day is generated by examining the current, or accumulated, count or time for each counter or timer.

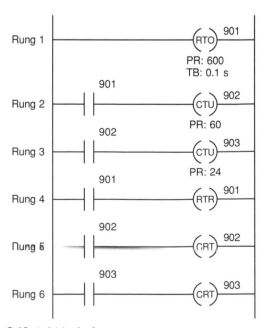

Fig. 8-13 A 24-h clock program.

The counter of rung 3 indicates the hour of day in 24-h military format. The current minutes are represented by the accumulated count value of the counter in rung 2. The timer of rung 1 displays the seconds of a minute as its current, or accumulated, time value.

The 24-h clock can be used to record the time of an event. Figure 8-14 illustrates the principle of this technique. In this application the time of the opening of a pressure switch is to be recorded. The circuit is set into operation by pressing the RESET button and setting the clock for the time of day. This starts the 24-hour clock and switches the SET indicating light on. Should the pres-

sure switch open at any time, the clock will automatically stop and the TRIP indicating light will switch on. The clock can then be read to determine the time of opening of the pressure switch.

8-5 INCREMENTAL ENCODER-COUNTER APPLICATIONS

The incremental encoder shown in Fig. 8-15 creates a series of square waves as its shaft is rotated. The encoder

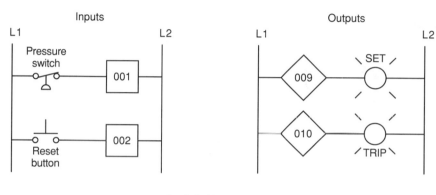

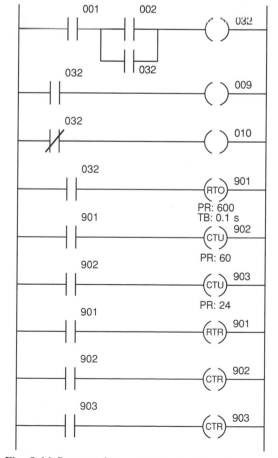

Fig. 8-14 Program for monitoring the time of an event.

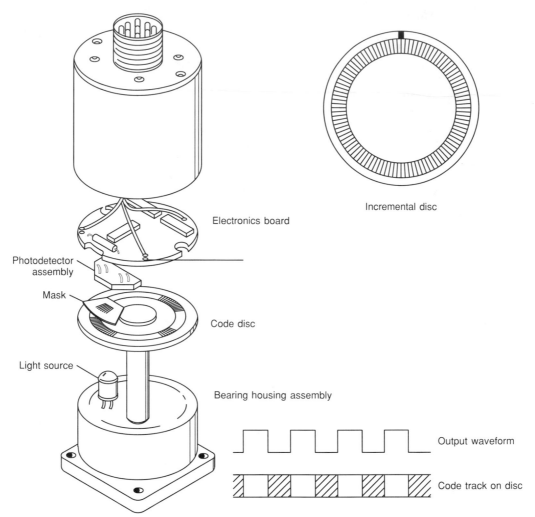

Fig. 8-15 Incremental encoder. *(Courtesy of BEI Motion Systems Company)*

disk interrupts the light as the encoder shaft is rotated to produce the square wave output waveform.

The number of square waves obtained from the output of the encoder can be made to correspond to the mechanical movement required. For example, to divide a shaft revolution into 100 parts, an encoder could be selected to supply 100 square wave cycles per revolution. By using a counter to count those cycles we could tell how far the shaft had rotated (Fig. 8-16).

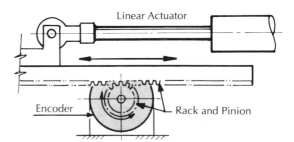

Fig. 8-16 Linear actuator *(Courtesy of BEI Motion Systems Company)*

REVIEW QUESTIONS

1. Name the three forms of PLC counter instructions and explain the basic operation of each.

2. State four pieces of information usually associated with a PLC counter instruction.

3. In a coil-formatted PLC counter instruction, what rule applies to the addressing of the CTU and CTR coils?

4. When is the output of a PLC counter energized?

5. When does the PLC counter instruction increment or decrement its current count?

6. The counter instructions of PLCs are normally retentive. Explain what this means

7. (a) Compare the operation of a standard PLC examine for ON contact with that of a transitional OFF-to-ON contact.

   (b) What is the normal function of a transitional contact used in conjunction with a counter?

PROBLEMS

1. Study the logic ladder program (Fig. 8-17) and answer the questions that follow:
   (a) What type of counter has been programmed?
   (b) What outputs are *real* and what outputs are *internal?*
   (c) What inputs are *real* and what inputs are *internal?*
   (d) When would output 009 be energized?
   (e) When would output 010 be energized?
   (f) Suppose your accumulated value is 024 and you lose ac line power to the controller. When power is restored to your controller, what will your accumulated value be?
   (g) Rung 4 goes TRUE and while it is TRUE rung 1 goes through five FALSE-to-TRUE transitions of rung conditions. What is the accumulated value of the counter after this sequence of events?
   (h) When will the count be incremented?
   (i) When will the count be reset?

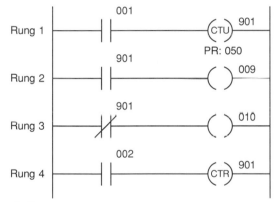

Fig. 8-17.

2. Study the logic ladder program (Fig. 8-18) and answer the questions that follow:
   (a) Suppose input 001 is switched from OFF to ON and remains on. How will the status of output 009 be affected?
   (b) Suppose input 001 is then switched from ON to OFF and remains off. How will the status of output 009 be affected?

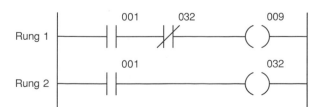

Fig. 8-18.

3. Study the logic ladder program (Fig. 8-19) and answer the questions that follow:
   (a) What type of counter has been programmed?
   (b) What input address will cause the counter to increment?
   (c) What input address will cause the counter to decrement?
   (d) What input address will reset the counter to a count of zero?
   (e) When would output 009 be energized?
   (f) Suppose the counter is first reset and then input 001 is actuated 15 times and input 002 is actuated 5 times. What is the accumulated count value?

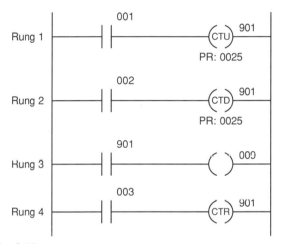

Fig. 8-19.

4. Design a PLC program and prepare a typical I/O connection diagram and logic ladder program for the following counter specifications:
   - Counts the number of times a push button is closed
   - Decrements the accumulated value of the counter each time a second push button is closed
   - Turns on a light any time the accumulated value of the counter is less than 20
   - Turns on a second light when the accumulated value of the counter is equal to or greater than 20
   - Resets the counter to zero when a selector switch is closed

5. Design a PLC program and prepare a typical I/O connection diagram and logic ladder program that will correctly execute the following control circuit:
   - Turns on a nonretentive timer when a switch is closed (preset value of timer is 10 s).
   - Timer is automatically reset by a programmed transitional contact when it times out.
   - Counter counts the number of times the timer goes to 10 s.
   - Counter is automatically reset by a second programmed transitional contact at a count of 5.
   - Latches on a light at the count of 5.

- Resets light to OFF when a selector switch is closed.

6. Design a PLC program and prepare a typical I/O connection diagram and logic ladder program that will correctly execute the following industrial control process (Fig. 8-20):

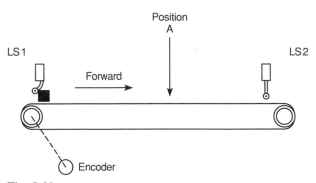

Fig. 8-20.

SEQUENCE OF OPERATION:

- Product in position (limit switch LS1 contacts close).
- START button is pressed and conveyor motor starts to move product forward toward position A (limit switch LS1 contacts close when the actuating arm returns to its normal position).
- Conveyor moves product forward to position A and stops (position detected by eight OFF-to-ON output pulses from the encoder, which are counted by an up-counter).
- A time delay of 10 s occurs, after which the conveyor starts to move the product to limit switch LS2 and stops (LS2 contacts close when actuating arm is hit by product).
- An emergency STOP button is used to stop the process at any time.
- If sequence is interrupted by an emergency stop, counter and timer are automatically reset.

9

PROGRAM CONTROL INSTRUCTIONS

Upon completion of this chapter you will be able to:

- Identify and list override and jump instructions
- Describe the function of immediate input and output instructions function
- Describe the forcing capability of the PLC
- Describe safety considerations built into PLCs and programmed into a PLC installation

---

9-1 MASTER CONTROL AND ZONE CONTROL INSTRUCTIONS

Several output-type instructions, which are often referred to as *override* instructions, provide a means of executing sections of the control logic if certain conditions are met. Instructions comprising the override instruction group include the *master control reset* (MCR), *zone control last (ZCL) state,* and JUMP (JMP) instructions. These operations are accomplished by using a series of conditional and unconditional branches and RETURN instructions. They all operate over a user-specified range, section, or zone of processor logic. The size of the zone is specified in some manner as part of the instruction.

The MCR instruction can be programmed to control an entire circuit or to control only selected rungs of a circuit. In the program of Fig. 9-1, the MCR is programmed to control an entire circuit. When the MCR instruction is FALSE, or de-energized, all *nonretentive* (nonlatched) rungs below the MCR will be *de-energized* even if the programmed logic for each rung is TRUE. All *retentive* rungs will remain in their *last state.* The MCR instruction establishes a zone in the user program in which all nonretentive outputs can be turned off simultaneously. Therefore, retentive instructions should not normally be placed within an MCR zone, because the

MCR zone maintains retentive instructions in the last active state when the instruction goes FALSE.

Figure 9-2 shows a programmed circuit with two MCR output instructions. By adding the second MCR instruction, the rungs between the two MCRs (rungs 2 to 7) are controlled. An MCR rung with conditional inputs (rung 2) is placed at the beginning of the fenced zone to be controlled. An MCR rung with *no* conditional inputs (rung 7) is placed at the end of the fenced zone to be controlled. When the MCR rung condition is TRUE, the referenced output is activated, and all rung outputs within the zone can be controlled by their respective input conditions.

If the MCR output is turned off, all nonretentive outputs within the fenced zone will be de-energized. Any rungs above or below the two MCR instructions will function normally, with the output being energized when the programmed logic is TRUE, *regardless* of whether the MCR is energized or not.

The ZCL instruction is similar to the MCR instruction. The differences between these two instructions are described by their respective names. The MCR (master control reset) will, when the zone is logic FALSE, reset all nonretentive outputs to the OFF state. The ZCL (zone control last state) will, when the zone is logic FALSE, leave all outputs in their last state.

Inputs   Outputs

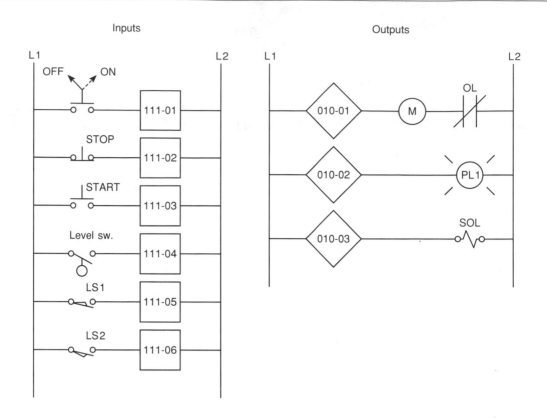

Logic ladder program

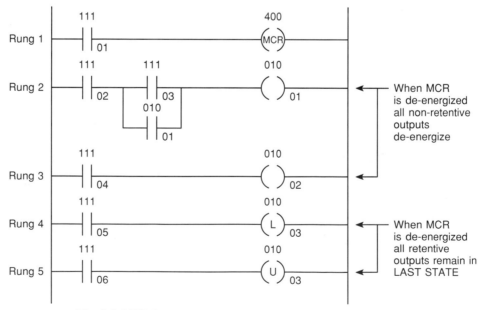

Fig. 9-1 MCR instruction programmed to control an entire circuit.

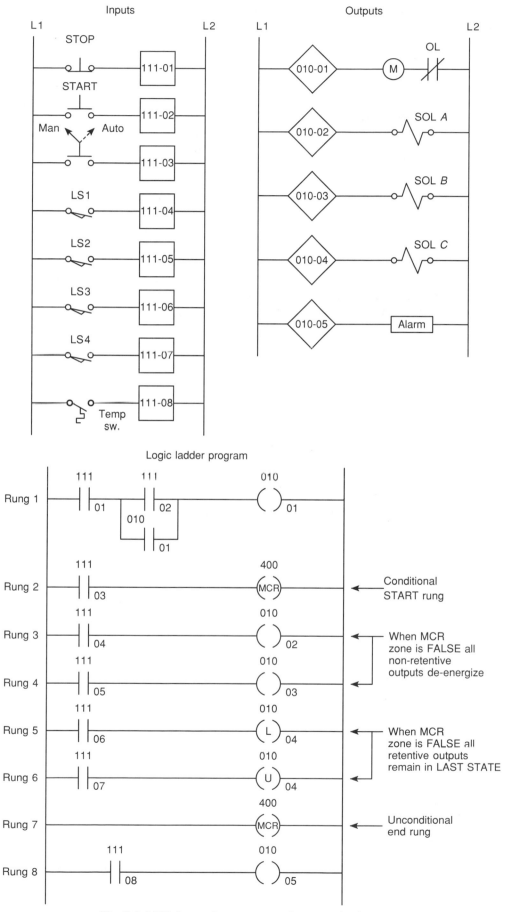

Fig. 9-2 MCR instruction programmed to control a fenced zone.

In Figure 9-3 you can see the MCR program of Fig. 9-2 programmed using the ZCL instruction in place of the MCR instruction. A ZCL output with conditional inputs is placed at the start of the fenced zone, and a ZCL output with no conditional inputs is placed at the end. If the EXAMINE ON instruction 111/03 is TRUE, the outputs within the zone are controlled by their respective rung input conditions. If the EXAMINE ON instruction 111/03 goes FALSE, the outputs within the zone will be held in their last state.

9-2 JUMP INSTRUCTIONS AND SUBROUTINES

As in computer programming, it is sometimes desirable to be able to jump over certain program instructions if certain conditions exist. The JUMP instruction is an output instruction used for this purpose. The advantage to the JUMP instruction is the ability to reduce the processor scan time by jumping over instructions not pertinent to the machine's operation at that instant.

Figure 9-4 shows a simple example of a JUMP-TO-LABEL program. The LABEL (LBL) instruction is used to identify the ladder rung that is the target destination of the JUMP instruction. The LABEL address number must match that of the JUMP instruction with which it is used. The LABEL instruction does not contribute to logic continuity, and for all practical purposes is always logically TRUE. When rung 4 has logic continuity, the processor is instructed to jump to rung 8 and continue to execute the main program from that point. Jumped rungs 5, 6, and 7 are not scanned by the processor. Input conditions are not examined and outputs that are controlled by these rungs remain in their last state. Any timers or counters programmed within the JUMP area cease to function and will not update themselves during this period. For this reason they should be programmed outside the jumped section in the main program zone.

Again as in computer programming, another valuable tool in PLC programming is to be able to escape from the main program and *go to* a program *subroutine* to perform certain functions and *then return* to the main program. Where a machine has a portion of its cycle that

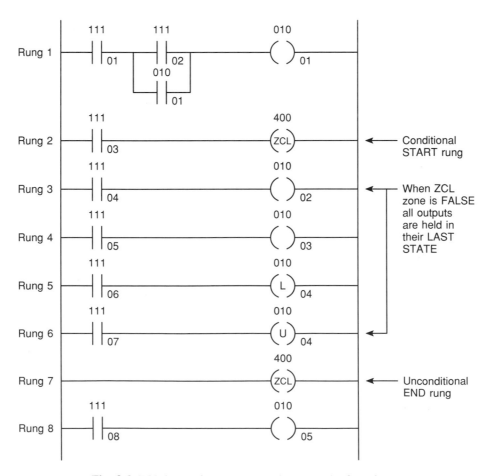

Fig. 9-3 ZCL instruction programmed to control a fenced zone.

must be repeated several times during one machine cycle, the subroutine can save a great deal of duplicate programming.

Figure 9-5 shows a simplified program that uses the JUMP-TO-SUBROUTINE (JSR), LABEL, and RETURN (RET) instructions. This program is similar to the one you will see later (see Chap. 11, p. 152) that is used for converting Celsius temperature to Fahrenheit. In the application programmed in Fig. 9-5 we want to record the converted temperature reading every 5 s. When the EXAMINE ON instruction 030/15 is TRUE, the processor jumps to the LABEL instruction 02 in the subroutine area and begins executing that subroutine. When the processor scan reaches the RETURN instruction, the RETURN instruction will send the scan back to the first instruction immediately following the JUMP-TO-SUBROUTINE instruction. The processor then continues to execute the remainder of the main program.

9-3 IMMEDIATE INPUT AND IMMEDIATE OUTPUT INSTRUCTIONS

The IMMEDIATE INPUT instruction (I) is a special version of the EXAMINE ON instruction used to read an input condition before the I/O update is performed. This operation interrupts the program scan when it is executed. After the IMMEDIATE INPUT instruction is executed, normal program scan resumes. This instruction is used with critical input devices that require updating in advance of the I/O scan.

The operation of the IMMEDIATE INPUT instruction

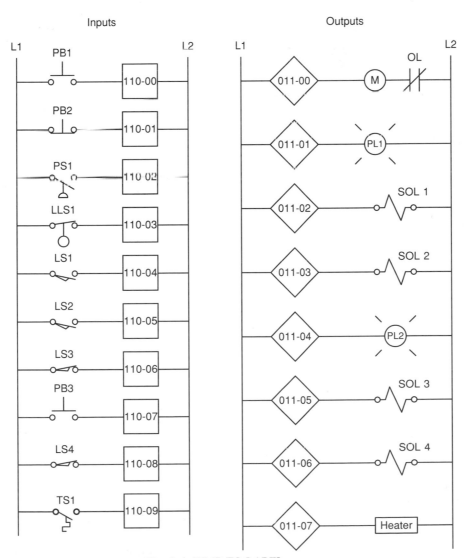

Fig. 9-4 JUMP-TO-LABEL program.

Logic ladder program

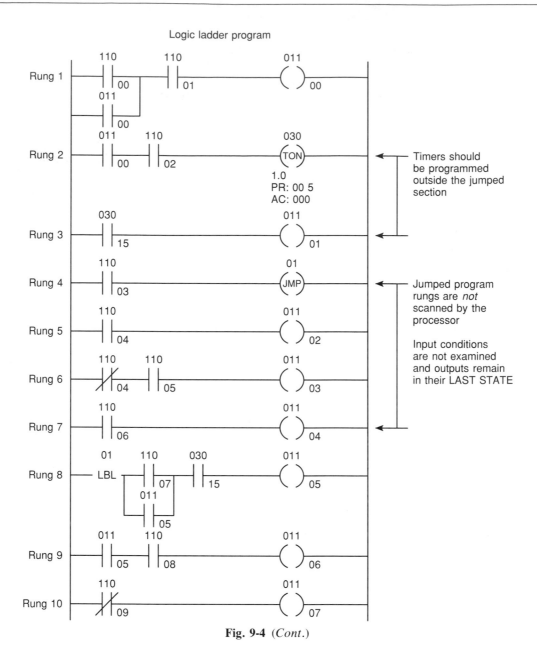

Fig. 9-4 (*Cont.*)

is illustrated in Fig. 9-6. When the program scan reaches the IMMEDIATE INPUT instruction, the scan is interrupted and the bits of the addressed word are updated. The IMMEDIATE INPUT is most useful if the instruction associated with the critical input device is at the middle or toward the end of the program. The IMMEDIATE INPUT is not needed near the beginning of the program, since the I/O scan has just occurred at that time. Although the IMMEDIATE INPUT instruction speeds the updating of bits, its scan-time interruption increases the total scan time of the program.

The IMMEDIATE OUTPUT (IOT) instruction is a special version of the OUTPUT ENERGIZE instruction used to update the status of an output device before the I/O

update is performed. The immediate output is used with critical output devices that require updating in advance of the I/O scan. The operation of the IMMEDIATE OUTPUT instruction is illustrated in Fig. 9-7. When the program scan reaches the IMMEDIATE OUTPUT instruction, the scan is interrupted and the bits of the addressed word are updated.

9-4   FORCING EXTERNAL I/O ADDRESSES

The forcing capability of a PLC allows the user to turn an external input or output on or off from the keyboard

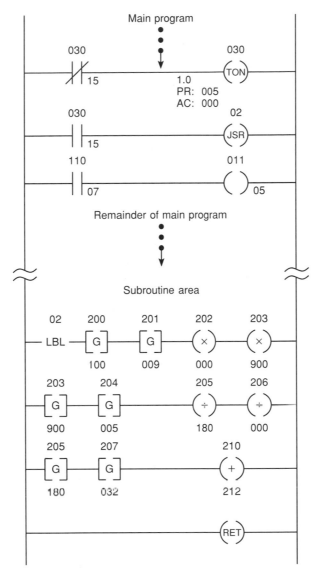

Fig. 9-5 JUMP-TO-SUBROUTINE and RETURN program.

of the programmer. This is accomplished regardless of the actual state of the field device (limit switch, etc.) for an input or logic rung for an output. This capability allows a machine or process to continue operation until a faulty field device can be repaired. It is also valuable during start-up and troubleshooting of a machine or process to simulate the action of portions of the program which have not yet been implemented.

When we force an input address, we are forcing the status bit of the instruction at the I/O address to an ON or OFF state. Figure 9-8 illustrates how an input is forced "on." The processor ignores the actual state of the limit switch (which is OFF) and considers input 003 as being in the ON state. The program scan records this, and the program is executed with this forced status. In other

words, the program is executed as if the limit switch were actually closed.

When we force an output address, we are forcing only the output terminal to an ON or OFF state. The status bit of the output instruction at the address is usually not affected. Figure 9-9 illustrates how an output is forced "on." The programming terminal acts in conjunction with the processor to turn output 005 "on" even though the output image table indicates that the user logic is setting the point to OFF. Output 006 remains off because the status bit of output 005 is not affected.

Care should be exercised when using forcing functions. Forcing functions should only be used by personnel who completely understand the circuit and the process machinery or driven equipment. An understanding of the

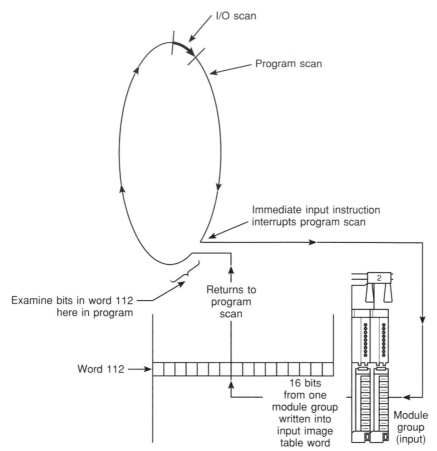

Fig. 9-6 IMMEDIATE INPUT instruction.

potential effect of forcing given inputs or outputs will have on machine operation is essential to avoid possible personal injury and equipment damage. Before using a force function, the user should check whether the "force" acts on the I/O point only, or whether it acts on the user logic as well as on the I/O point. Most programming terminals provide some visible means of alerting the user that a force is in effect.

9-5 SAFETY CIRCUITRY

Sufficient emergency circuits must be provided to stop either partially or totally the operation of the controller or the controlled machine or process. These circuits should be hard-wired outside the controller, so that in the event of total controller failure, independent and rapid shutdown means are available.

Figure 9-10 shows a typical safety wiring diagram for a PLC installation. A main disconnect switch is installed on the incoming power lines as a means of removing power from the entire programmable controller system. This is followed by an isolation transformer, which is used to isolate the controller from the main power distribution system and step the voltage down to 120 V ac. A hard-wired master control relay or contactor circuit is included to control the ac power to the processor and the input and output devices. This circuit provides connection for mushroom-head emergency STOP switches, emergency pull-rope switches, and end-of-travel limit switches strategically placed for maximum safety. When any of these emergency switches are opened, power to input and output devices is removed. Power continues to be supplied to the controller power supply so that any diagnostic indicators on the processor module can still be observed. Note that the master control relay is *not* a substitute for a disconnect switch. When replacing any module, replacing output fuses, or working on equipment, the main disconnect switch should be pulled and locked out.

Certain safety considerations should be developed as

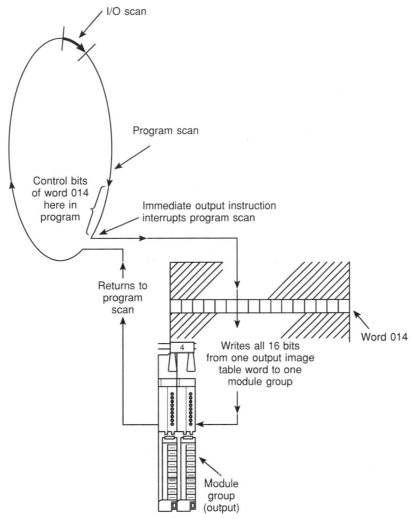

Fig. 9-7 IMMEDIATE OUTPUT instruction.

part of the PLC program. A PLC program for any application will be only as safe as the time and thought spent on both personal and hardware considerations make it. One such consideration involves the use of a motor starter *seal-in* contact (Fig. 9-11) in place of the programmed contact referenced to the output coil instruction. The use of the field-generated starter auxiliary contact status in the program is more costly in terms of field wiring and hardware, but is *safer* because it provides positive feedback to the processor as to the exact status of the motor. Assume for example that the OL contact of the starter opens under an overload condition. The motor, of course, would stop operating since power would be lost to the starter coil. If the program was written using an NO contact, instruction referenced to the output coil instruction as the seal-in for the circuit, the processor would never know power has been lost to the motor. When the OL was reset, the motor would restart instantly, creating a potentially unsafe operating condition.

Another safety consideration concerns the wiring of STOP buttons. A STOP button is generally considered a safety function as well as an operating function. As such *it should be wired using an NC contact and programmed to examine for an ON condition*. Using an NO contact programmed to examine for an OFF condition (Fig 9-12) will produce the same logic but is not considered to be as safe. Assume the latter configuration is used. If, by some chain of events, the circuit between the button and the input point were to be broken, the STOP button could be depressed forever, but the PLC logic could never react to the STOP command since the input would never be TRUE. The same holds true if power were lost to the STOP button control circuit. If the NC wiring configuration is used, the input point receives power continuously unless the STOP function is desired. Any faults occurring with the STOP circuit wiring, or a loss of circuit power, would effectively be equivalent to an intentional STOP.

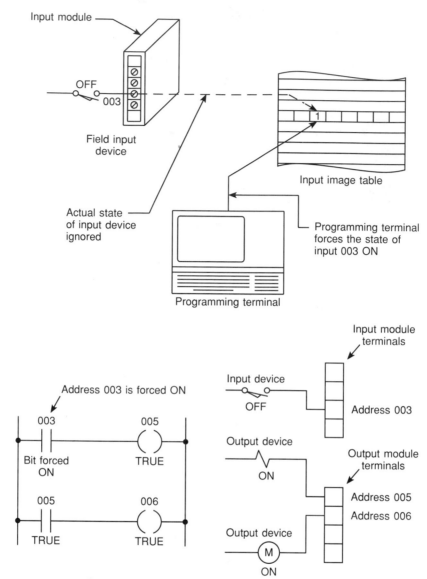

Fig. 9-8 Forcing an input address ''on.''

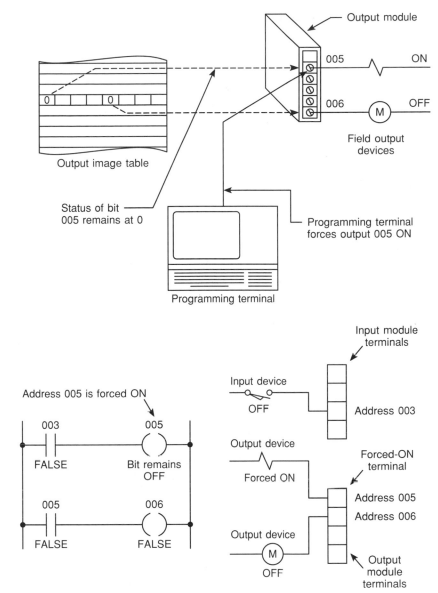

Fig. 9-9 Forcing an output address "on."

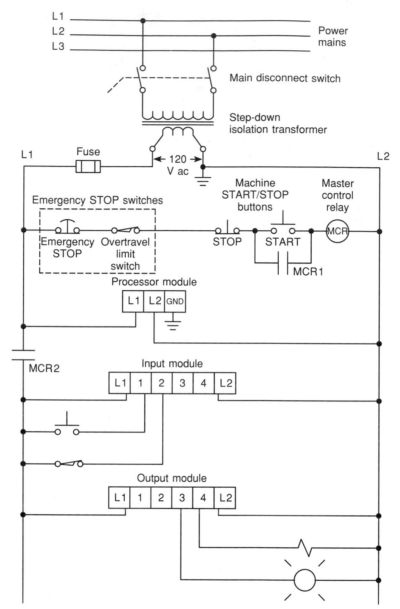

Fig. 9-10 Typical PLC safety wiring diagram.

Inputs

Output

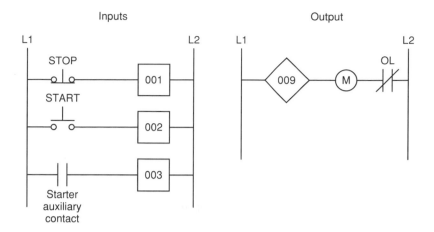

Logic ladder program

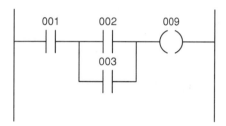

Fig. 9-11 Motor starter program using the auxiliary contact. *(Photo courtesy of Square D Company)*

Inputs   Output

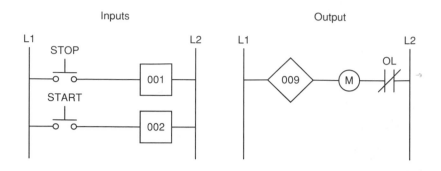

Logic ladder program

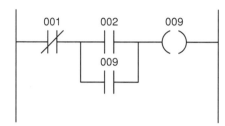

Fig. 9-12 NO stop configuration.

REVIEW QUESTIONS

1. (a) Two MCR output instructions are to be programmed to control a section of a program. Explain the programming procedure to be followed.
   (b) State how the status of the output devices within the fenced zone will be affected when the MCR instruction makes a FALSE-to-TRUE transition.
   (c) State how the status of the output devices within the fenced zone will be affected when the MCR instruction makes a TRUE-to-FALSE transition.

2. Explain the difference between the MCR and ZCL program control instructions.

3. What is the main advantage of the JUMP instruction?

4. What type of instructions are not normally included inside the jumped section of a program? Why?

5. (a) What is the purpose of the LABEL instruction in the JUMP-TO-LABEL instruction pair?
   (b) When the JUMP-TO-LABEL instruction is executed, in what way are the jumped rungs affected?

6. (a) Explain what the JUMP-TO-SUBROUTINE instruction allows the program to do.
   (b) In what type of machine operation can this instruction save a great deal of duplicate programming?

7. (a) When are the IMMEDIATE INPUT and IMMEDIATE OUTPUT instructions used?
   (b) Why is it of little benefit to program an IMMEDIATE INPUT or OUTPUT instruction near the beginning of a program?

8. (a) What does the forcing capability of a PLC allow the user to do?
   (b) Outline two practical uses for forcing functions.
   (c) Why should extreme care be exercised when using forcing functions?

9. Why should emergency stop circuits be hard-wired instead of programmed?

10. State the function of each of the following in the basic safety wiring for a PLC installation:
    (a) Main disconnect switch
    (b) Isolation transformer
    (c) Emergency STOPS
    (d) Master control relay

11. When programming a motor starter circuit, why is it safer to use the starter seal-in auxiliary contact in place of a programmed contact referenced to the output coil instruction?

12. When programming STOP buttons, why is it safer to use an NC button programmed to examine for an ON condition than an NO button programmed to examine for an OFF condition?

PROBLEMS

1. Answer the questions, in sequence, for the MCR program (Fig. 9-13), assuming the program has just been entered and the PLC placed in the RUN mode with all switches turned off.

   (a) Switches 002 and 003 are turned on. Will outputs 009 and 010 come on? Why?

   (b) With switches 002 and 003 still on, switch 001 is turned on. Will output 009 or 010 or both come on? Why?

   (c) With switches 002 and 003 still on, switch 001 is turned off. Will both outputs 009 and 010 de-energize? Why?

   (d) With all other switches off, switch 006 is turned on. Will the timer time? Why?

   (e) With switch 006 still on, switch 005 is turned on. Will the timer time? Why?

   (f) With switch 006 still on, switch 005 is turned off. What happens to the timer? If the timer was an RTO type instead of a TON, what would happen to the accumulated value?

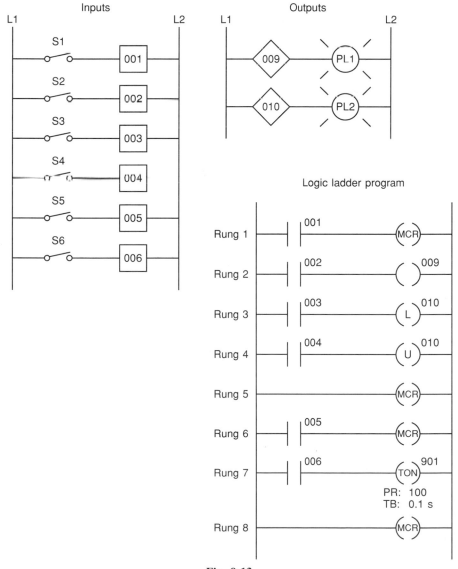

Fig. 9-13

2. Answer the questions, in sequence, for the ZCL program (Fig. 9-14), assuming that the program has just been entered and the PLC placed in the RUN mode with all switches turned off.

(a) Switches 002 and 005 are turned on. Will any outputs come on? Why?

(b) With switches 002 and 005 still on, switch 001 is turned on. What will happen to outputs 009 and 010? Why?

(c) With switches 002 and 005 still on, switch 001 is turned off. What will happen to the outputs?

(d) Switch 002 is now turned off. Will output 009 go off? Why?

(e) With all other switches off, switch 004 is turned on. Will timer 901 start timing? Why?

(f) With switch 004 still on, switch 003 is turned on. will the timer function correctly?

(g) With the timer timing, switch 003 is turned off. What will happen to the timer's accumulated value?

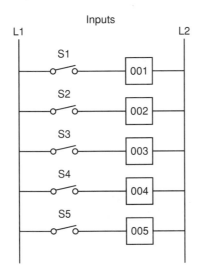

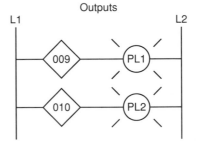

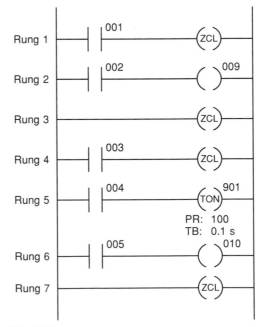

Fig. 9-14

3. Answer the questions, in sequence, for the JUMP-TO-LABEL program (Fig. 9-15). Assume all switches are turned *off after each operation.*
   (a) Switch 112/02 is turned on. Will output 013/00 be energized? Why?
   (b) Switch 112/01 is turned on *first,* then switch 112/04 is turned on. Will output 013/04 be energized? Why?
   (c) Switch 112/02 is turned on and output 013/00 is energized. Next, switch 112/01 is turned on. Will output 013/00 be energized or de-energized after turning on switch 112/01? Why?
   (d) All switches are turned on in order according to the following sequence: 112/00, 112/01, 112/02, 112/04, 112/03. Which pilot lights will turn on?

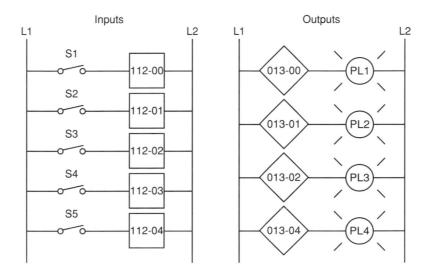

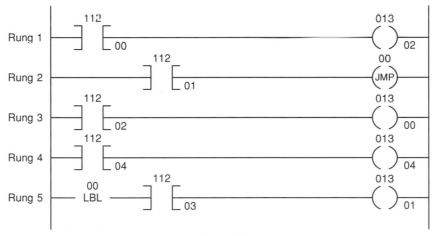

Fig. 9-15

4. Answer the questions, in sequence, for the jump-to-subroutine and return program (Fig. 9-16). Assume all switches are turned *off after each operation*.
   (a) Switches 112/00, 112/02, 112/10, and 112/11 are all turned on. Which pilot light will *not* be turned on? Why?
   (b) Switch 112/01 is turned on and then switch 112/10 is turned on. Will output 013/10 be energized? Why?

(c) To what rung does the RET instruction return the program scan to?
(d) Is rung 6 part of the subroutine area? Why?
(e) Assuming all switches are turned on, in what order will the rungs be scanned?
(f) Assuming all switches are turned off, in what order will the rungs be scanned?

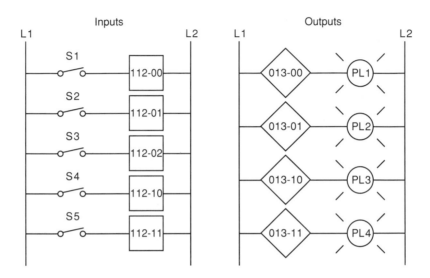

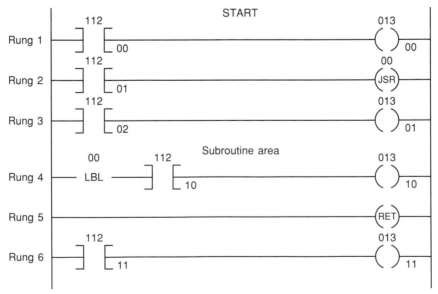

Fig. 9-16

5. Answer the questions, in sequence, for Fig. 9-17. Assume all switches are turned *off after each operation*.

    (a) Switches 112/01, 112/14, and 112/04 are turned on in order. Will output 013/16 be energized? Why?

    (b) All switches, except 112/07, are turned off. Will RTO/31 start timing? Why?

    (c) Switches 112/02 and 112/10 are turned on in order. Will pilot light PL2 come on? Why?

    (d) When will timer TON/030 function?

    (e) Assuming all switches are turned on, in what order will the rungs be scanned?

    (f) Assuming all switches are turned off, in what order will the rungs be scanned?

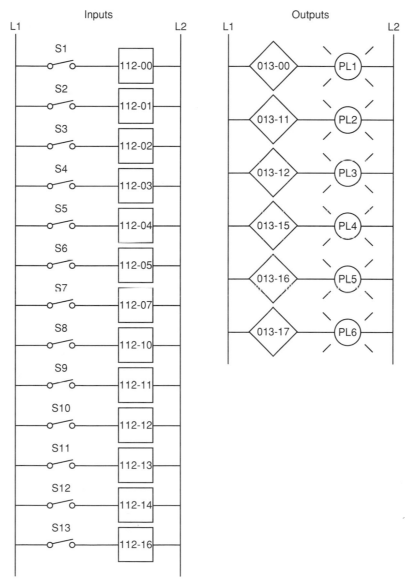

Fig. 9-17

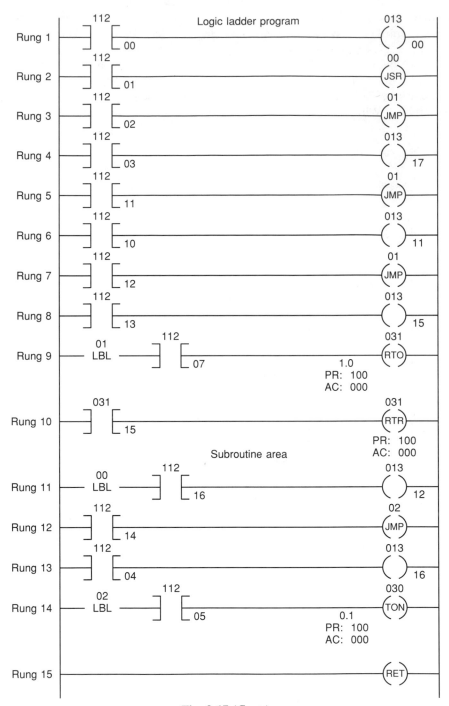

Fig. 9-17 (*Cont.*)

10

DATA MANIPULATION INSTRUCTIONS

Upon completion of this chapter you will be able to:

- Define data manipulation and apply it by writing a PLC program
- Interpret data transfer and data compare instructions as they apply to a PLC program
- Compare the operation of discrete I/Os with that of multibit and analog types
- Describe the basic operation of a closed-loop control system

10-1 DATA MANIPULATION

Data manipulation instructions enable the programmable controller to take on some of the qualities of a computer system. Many PLC's now have this ability to manipulate data that is stored in memory. It is this extra computer characteristic that gives the PLC capabilities that go far beyond the conventional relay equivalent instructions.

Each data manipulation instruction requires two or more words of data memory for operation. The words of data memory in singular form may be referred to either as *registers* or as *words*, depending on the manufacturer. The terms *table* or *file* are generally used when a *consecutive* group of data memory words is being referred to. Figure 10-1 illustrates the difference between a word and file. The data contained in files and words will be in

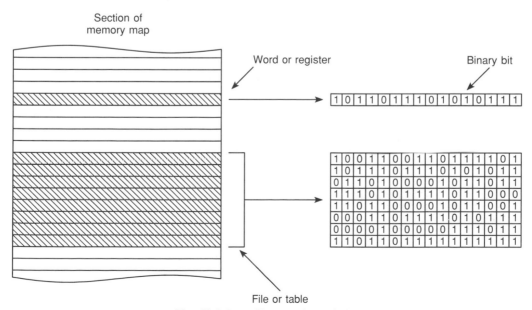

Fig. 10-1 Data files, words, and bits.

the form of binary *bits* represented as series of 1's and 0's.

The data manipulation instructions allow the movement, manipulation, or storage of data in either single- or multiple-word groups from one data memory area of the PLC to another. Use of these PLC instructions in applications that require the generation and manipulation of large quantities of data greatly reduces the complexity and quantity of the programming required. Data manipulation can be placed in two broad categories: *data transfer* and *data comparison*.

To simplify the explanation of the various data manipulation instructions available, the instruction protocol for the Allen-Bradley PLC-2 family of PLCs will be used. Again, even though the format and instructions vary with each manufacturer, the concepts of data manipulation remain the same.

10-2 DATA TRANSFER OPERATIONS

Data transfer instructions simply involve the transfer of the contents from one word or register to another. Figure 10-2a and b illustrate the concept of moving numerical binary data from one memory location to another. Figure 10-2a shows that numerical data is stored in word 130 and that no information is currently stored in word 040. Figure 10-2b shows that after the data transfer has oc-

curred, word 040 now holds the exact, or duplicate, information that is in word 130. If word 040 had other information already stored (rather than all 0's), this information would have been replaced. When new data replaces existing data in this manner, the process is referred to as *writing over the existing data*.

Data transfer instructions can address almost any location in the memory. Prestored values can be automatically retrieved and placed in any new location. That location may be the preset register for a timer or counter or even an output register that controls a seven-segment display.

There are two data transfer instructions: GET and PUT. GET instructions tell the processor to go *get* a value stored in some word. The GET instruction is programmed in the *condition* portion of a logic rung. It is always a logic TRUE instruction that will get an entire 16-bit word from a specified location in the data table. Figure 10-3 shows a typical GET instruction, which tells the processor to get the value 005 that is stored in word 020.

PUT instructions tell the processor where to *put* the information it obtained from the GET instruction. The PUT instruction is programmed in the *output* portion of the logic rung. A PUT instruction receives all 16 bits of data from the immediately preceding GET instruction. It is used to store the result of other operations in the memory location (word or register) specified by the PUT coil.

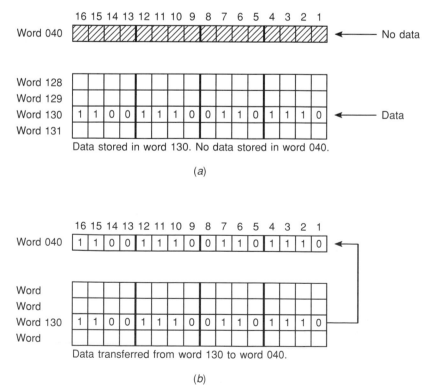

(a)

(b)

Fig. 10-2 Data transfer concept.

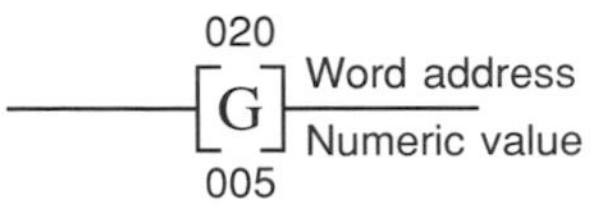

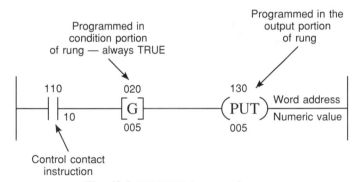

Fig. 10-3 Allen-Bradley GET instruction.

Figure 10-4 shows a typical PUT instruction, which tells the processor to put the information obtained from the GET instruction into word 130.

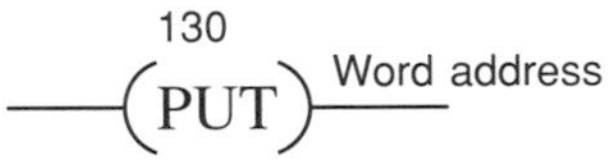

Fig. 10-4 Allen-Bradley PUT instruction.

The PUT instruction is used with the GET instruction to form a data transfer rung. Figure 10-5 shows an example of a data transfer rung. When input 110/10 is TRUE, the GET/PUT instructions tell the processor to get the numeric value 005 stored in word 020 and put it into word 130. In every case the PUT instruction must be preceded by a GET instruction.

Figure 10-6 shows an example of how the GET/PUT data transfer instructions can be used to change the preset time of an on-delay timer. In this example, delay times of 5 or 10 s are selected by means of selector switch SS1. When the selector switch is in the 10-s position, rung 2 has logic continuity and rung 3 does not. As a

Fig. 10-5 GET/PUT data transfer rung.

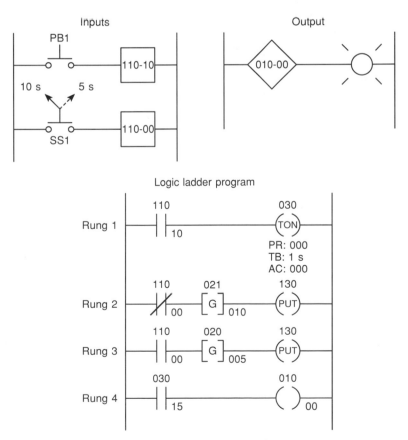

Fig. 10-6 Changing the preset value of a timer with GET/PUT instructions.

result, the processor is told to get the value of 010 stored in word 021 and put it into word 130. Address 130 is where the preset value of the on-delay timer 030 is stored. Therefore, the preset value of the timer 030 will change from 000 to 010. When push button PB1 is closed, there will be a 10-s delay period before output 010/00 is energized. The timer contact addressed 030/15 is the *done* bit of the timer. Its NO contact closes when the preset value is equal to the accumulated value, thus indicating that the timer instruction has completed its function. To change the preset value of the timer to 5 s, the selector switch is turned to the 5-s position. This makes rung 3 TRUE and rung 2 FALSE. As a result, the preset value of the timer 030 will change from 010 to 005. Pressing push button PB1 closed will now result in a 5-s time-delay period before output 010/00 is energized.

Figure 10-7 shows an example of how the GET/PUT data transfer instruction can be used to change the preset

count of an up-counter. In this example, a limit switch programmed to operate a counter counts the products coming off a conveyor line onto a storage rack. Three different types of products are run on this line. The storage rack has room for only 300 boxes of product A or 175 boxes of product B or 50 boxes of product C. Three switches are provided to select the desired preset counter value depending on the product line—A, B, or C—being manufactured. A RESET button is provided to reset the accumulated count to zero. A pilot lamp is switched on to indicate when the storage rack is full.

10-3 DATA COMPARE INSTRUCTIONS

Data compare instructions compare the data stored in two or more words (or registers) and make decisions based on the program instructions. Compare instructions oper-

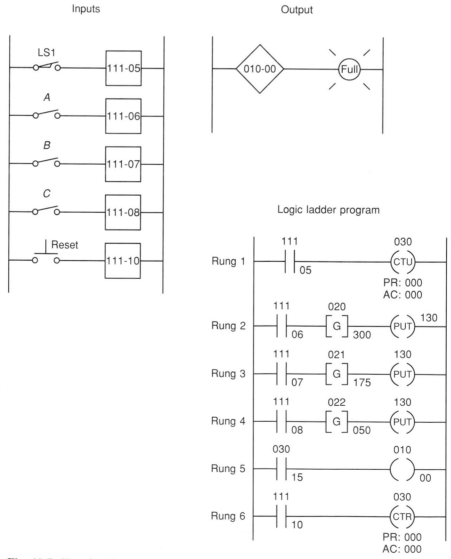

Fig. 10-7 Changing the preset value of a counter with a GET/PUT instruction.

ate in conjunction with GET instructions. Like GET instructions, compare instructions are programmed in the condition portion of a logic rung. Numeric values in two words of memory can be compared for each of the following conditions depending on the PLC:

NAME	SYMBOL
LESS THAN	(<)
EQUAL TO	(=)
GREATER THAN	(>)
LESS THAN OR EQUAL TO	(≤)
GREATER THAN OR EQUAL TO	(≥)

Data comparison concepts have already been used with the timer and counter instructions of previous units. In both of these instructions an output was turned on or off when the accumulated value of the timer or counter equaled its preset value (AC = PR). What actually occurred was that the accumulated numeric data in one memory word was *compared* to the reset value of another memory word on each scan of the processor. When the processor saw that the accumulated value was equal to (=) the preset value, it switched the output on or off.

Figure 10-8 shows a logic rung that uses an EQUAL TO (=) instruction. A GET (G) instruction is always programmed preceding any data compare instruction. The EQUAL TO instruction will have a logic TRUE condition only when the value stored in the EQUAL TO instruction is the *same as* the value stored in the GET instruction. In this example the GET value is the changing variable and is compared to the reference value of the EQUAL TO instruction for an EQUAL TO condition. If input 120/03 is TRUE, when the GET value YYY equals 100 the comparison is TRUE and logic rung continuity is established.

Figure 10-9 shows a logic rung that uses a LESS THAN (<) instruction. The LESS THAN instruction will have a logic TRUE condition only when the value stored in the GET instruction is *less than* the value stored in the LESS THAN instruction. In this example if input 120/03 is TRUE, rung continuity will be established any time the GET value is *less than* 100.

A GREATER THAN (>) comparison is also made with the GET/LESS THAN pair of instructions. This time the GET *instruction* value is the *reference* and the LESS THAN instruction value is the changing variable. Figure 10-10 shows the program required to implement a GREATER THAN (>) comparison. In this example, if input 120/03 is TRUE, rung continuity will be established any time the GET value is *greater than* 100.

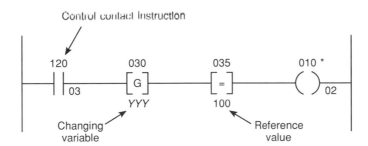

* Output 010-02 will be energized when input 120-03 is TRUE and the value in word 030 is *equal to* the reference value *100* in word 035.

Fig. 10-8 GET/EQUAL TO logic rung.

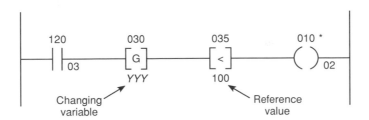

* Output 010-02 will be energized when input 120-03 is TRUE and the value in word 030 is *less than* the reference value *100* in word 035.

Fig. 10-9 GET/LESS THAN logic rung.

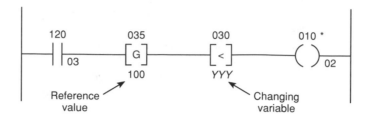

* Output 010-02 will be energized when input 120-03
  is TRUE and the value in word 030 is *greater than*
  the reference value *100* in word 035.

Fig. 10-10 GET/GREATER THAN logic rung.

For a LESS THAN OR EQUAL TO (≤) comparison, the rung is programmed with one GET instruction followed by LESS THAN (<) and EQUAL TO (=) instructions in parallel. The GET value is the changing value. The LESS THAN and EQUAL TO instructions are assigned a reference value. Figure 10–11 shows the program required to implement a LESS THAN OR EQUAL TO (≤) comparison. In this example, if input 120/03 is TRUE, rung continuity will be established any time the GET value is either *less than* or *equal to* 100.

Note that both the LESS THAN (<) and EQUAL TO (=) instructions use the same word address and have the same reference value.

The GREATER THAN OR EQUAL TO (≥) comparison rung is also programmed with one GET instruction followed by LESS THAN and EQUAL TO instructions connected in parallel. In this instance however, the *GET* instruction value is the *reference value*. The LESS THAN and EQUAL TO values change, and are compared to the GET value. Figure 10-12 shows the

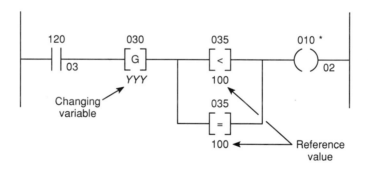

* Output 010-02 will be energized when input 120-03
  is TRUE and the value in word 030 is *less than* or
  equal to the reference value *100* in word 035.

Fig. 10-11 GET/LESS THAN OR EQUAL TO logic rung.

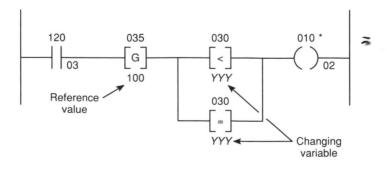

* Output 010-02 will be energized when input 120-03
  is TRUE and the value in word 030 is *greater than* or
  equal to the reference value *100* in word 035.

Fig. 10-12 GET/GREATER THAN OR EQUAL TO logic rung.

program required to implement a GREATER THAN OR EQUAL TO ($\geq$) comparison. In this example, if input 120/03 is TRUE, rung continuity will be established any time the LESS THAN OR EQUAL TO values are *greater than* or *equal to* 100.

10-4 DATA MANIPULATION PROGRAMS

As mentioned, data manipulation instructions give new dimension and flexibility to the programming of control circuits. For example, consider the original relay-operated time-delay circuit of Fig. 10-13. This circuit uses three pneumatic time-delay relays to control four solenoid valves. When the START push button is pressed, solenoid A is energized immediately, solenoid B is energized 5 s later, solenoid C is energized 10 s later, and solenoid D is energized 15 s later.

This circuit could be implemented using a conventional PLC program and three internal timers. However, the same circuit can be programmed using only *one* internal timer along with *data compare* statements, and this will result in a savings of memory words. Figure 10-14 shows the program required to implement the circuit using only one internal timer. Assuming the STOP button (111/05) is closed, when the START button (111/06) is pushed, output 010/01 will energize. As a result, solenoid A will switch on; contact 010/01 will close to seal in output 010/01; contact 010/01 will close to start on-delay timer TON 030. The timer has been preset to 15 s, and the accumulated time will be stored in word 030. Output 010/04 will energize (through the *done* bit 030/15) after

a total time delay of 15 s to energize solenoid D. Output 010/02 will energize after a total time delay of 5 s when the accumulated time becomes equal to and then greater than the GET reference time (020/005) of 5 s. This, in turn, will energize solenoid B. Output 010/03 will energize after a total time delay of 10 s when the accumulated time becomes equal to and then greater than the GET reference time (021/010) of 10 s. This, in turn, will energize solenoid C.

Figure 10-15 shows an on-delay timer program that uses the EQUAL TO instruction. When the switch (111/05) is closed, timer 031 will begin timing. Both GET instructions are addressed to get the accumulated value from the timer while it is running. The EQUAL TO instruction in rung 2 has the BCD value of 005 stored in address 022. When the accumulated value of the timer is equal to 005, the EQUAL TO instruction in rung 2 will become logic TRUE for 1 s. As a result, *latch* output 010/01 will energize to switch the light on. Then, when the accumulated value of the timer reaches 015, the EQUAL TO instruction (023/015) in rung 3 will be TRUE for 1 s. As a result, *unlatch* output 010/01 will energize to switch the light off. Therefore, when the switch is closed, the light will come on after 5 s, stay on for 10 s, and then turn off.

Figure 10-16 shows an up-counter program that uses the LESS THAN instruction. Up-counter 031 of rung 1 will increment by 1 for every FALSE-to-TRUE transition of push-button input 111-05. Rung 2 contains a GET instruction that is addressed to the accumulated value of the counter (031) and a LESS THAN instruction that has the BCD value 020 stored in address 022. The LESS THAN instruction will be TRUE as long as the value contained in the GET instruction is less than the value

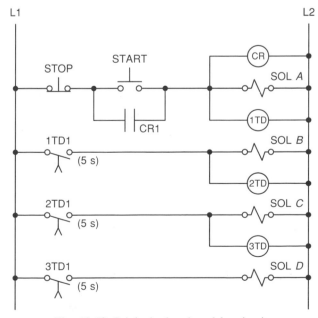

Fig. 10-13 Original relay time-delay circuits.

Inputs

Outputs

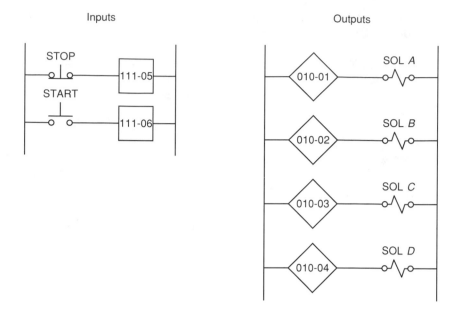

Logic ladder program

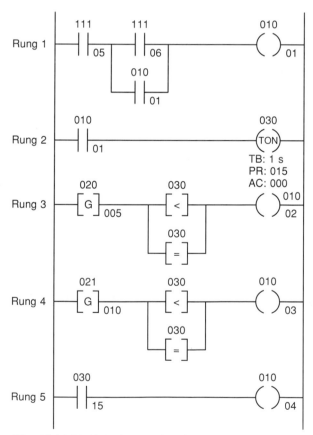

Fig. 10-14 Multiple timers using data compare statements.

stored in the LESS THAN instruction. Therefore, the output 010/01 and the light will be on when the accumulated value of the counter is between 000 and 019. As soon as the counter's accumulated value is 020, the LESS THAN instruction will go FALSE, turning off output 010/

01 and the light. When the counter's accumulated value reaches its preset value of 050, the counter reset will be energized through the done bit 15 of word 031 to reset the accumulated count to zero.

The program of Fig. 10-16 can be modified as shown

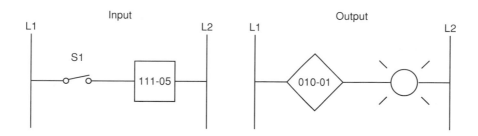

Logic ladder program

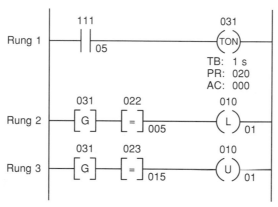

Fig. 10-15 GET/EQUAL TO timer program.

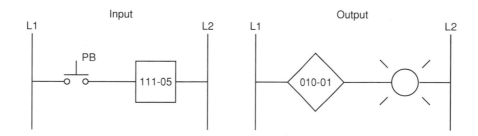

Logic ladder program

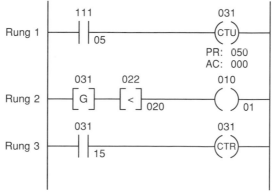

Fig. 10-16 GET/LESS THAN counter program.

in Fig. 10-17 to achieve a GREATER THAN instruction. In this instance, the GET instruction has a fixed reference value of 020 stored in address 022, and the LESS THAN instruction is addressed to the counter's accumulated value. Output 010/01 and the light will now be on when the counter's accumulated value is from 021 to 050, at which point rung 3 will reset the counter.

The use of comparison instructions is generally straightforward. However, one common programming error involves the use of these instructions in a PLC program to control the flow of a raw material into a vessel. The receiving vessel has its weight continuously monitored by the PLC program as it fills. When the weight reaches a preset value, the flow is cut off. While the vessel fills, the PLC performs a comparison between the vessel's current weight and a predetermined constant programmed in the processor. If the programmer uses only the EQUAL TO instruction, problems may result. As the vessel fills, the comparison for equality will be FALSE. At the instant the vessel weight reaches the desired preset value of the EQUAL TO instruction, the instruction becomes TRUE and the flow is stopped. However, should the supply system leak additional material into the vessel, the total weight of the material could rise *above* the preset value, causing the instruction to go FALSE and the vessel to overfill. The simplest solution to this problem is to program the comparison instruction as GREATER THAN OR EQUAL TO. This way any excess material entering the vessel will not affect the filling operation. It may be necessary, however, to include additional programming to indicate a serious overfill condition.

10-5   NUMERICAL DATA I/O INTERFACES

The expanding data manipulation processing capabilities of PLCs led to a new class of I/O interfaces known as numerical data I/O interfaces. In general, numerical data I/O interfaces can be divided into two groups: those that provide interface to *multibit* digital devices and those that provide interface to *analog* devices.

The multibit digital devices are like the discrete I/O in that processed signals are discrete (ON/OFF). The difference is that with the discrete I/O only a *single* bit is required to read an input or control an output. Multibit interfaces allow a *group* of bits to be input or output *as a unit*. They are used to accommodate devices that require BCD inputs or outputs.

Figure 10-18 shows a BCD input interface (register input module) connected to *thumbwheel switches* (TWS). This interface is used to input data into specific register or word locations in memory to be used by the control program. Register input modules generally accept voltages in the range of 5 V dc (TTL) to 24 V dc. They are grouped in a module containing 16 or 32 inputs, corresponding to one or two I/O registers, respectively. Data manipulation instructions such as GET are used to access the data from the register input interface.

Figure 10-19 shows a BCD output interface (register output module) connected to a seven-segment LED display board. This interface is used to output data from a specific register or word location in memory. Register output modules generally provide voltages that range from 5 V dc (TTL) to 30 V dc and have 16 or 32 output lines, corresponding to one or two I/O registers, respectively. The BCD register output module can also be used to drive small dc loads that have current requirements in the 0.5-A range.

Figure 10-20 shows a PLC program that uses a BCD input interface module connected to thumbwheel switches and a BCD output interface module connected to an LED display board. The decimal setting of the thumbwheel switches is monitored by the LED display board. In this program, the setting of the thumbwheel switches is compared to that of the reference number (100) stored in word 035. Output 011/12 will be energized when input 110/02 is TRUE and the value of the thumbwheel switches is equal to the reference value 100 in word 035.

An analog I/O will allow monitoring and control of analog voltages and currents. Figure 10-21 (see page 142) shows how an analog input interface operates. The analog input module contains the circuitry necessary to accept analog voltage or current signals from field devices. These voltage or current inputs are converted from an analog to a digital value by an A/D converter circuit. The conversion value, which is proportional to the analog signal, is passed through the controller's data bus and stored in a specific register or word location in memory for later use by the control program.

An analog output interface module receives numerical data from the processor; this data is then translated into a proportional voltage or current to control an analog

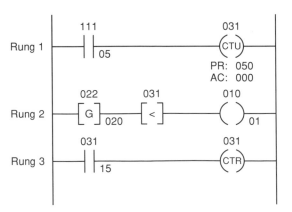

Fig. 10-17 GET/GREATER THAN counter program.

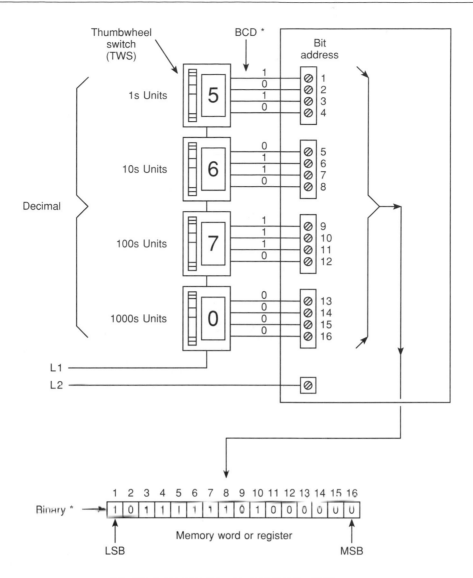

* In this illustration, it is assumed that the PLC processor is programmed to convert the BCD value into an equivalent binary value, and then to load that binary value into the register.

Fig. 10-18 BCD input interface module.

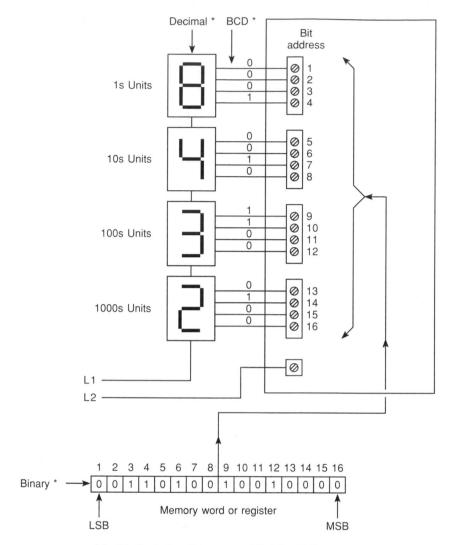

* In this illustration, it is assumed that the PLC
processor is programmed to convert the binary value
in the register into an equivalent BCD value, and
that the LED display board is responsible for
encoding the programmable controller BCD output
to produce the correct decimal digit on each display.

Fig. 10-19 BCD output interface module.

Inputs

Outputs

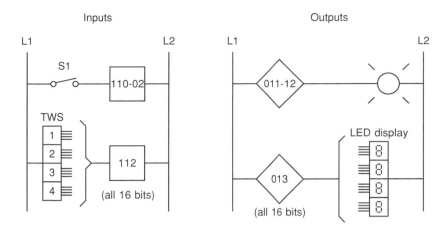

Logic ladder program

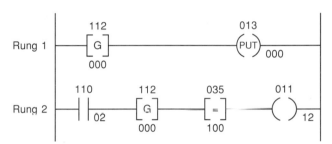

Fig. 10-20 BCD I/O program.

field device. Figure 10-22 shows how an analog output interface operates. Data from a specific register or word location in memory is passed through the controller's data bus to a D/A converter. The analog output from the converter is then used to control the analog output device. These output interfaces normally require an external power supply that meets certain current and voltage requirements.

10-6 SETPOINT CONTROL

Setpoint control in its simplest form compares an input value, such as an analog or thumbwheel inputs, to a setpoint value. A discrete output signal is provided if the input value is less than, equal to, or greater than the setpoint value.

The temperature control program of Fig 10-23 (see page 144) is one example of setpoint control. In this application, a PLC is to provide for simple OFF/ON control of the electric heating elements of an oven. The oven is to maintain an average setpoint temperature of 600°F, with a variation of about 1 percent between the OFF and ON cycles. Therefore, the electric heaters will be turned

on when the temperature of the oven is 597°F or less and stay on until the temperature rises to 603°F or more. The electric heaters stay off until the temperature drops down to 597°F, at which time the cycle repeats itself. Rung 2 contains a GET/LESS OR EQUAL TO logic instruction and rung 3 contains the equivalent of a GET/GREATER THAN OR EQUAL TO logic instruction. Rung 4 contains the logic for switching the heaters on and off according to the high and low setpoints. Rung 1 contains the logic that allows the thermocouple temperature to be monitored by the LED display board.

Several common setpoint control schemes can be performed by different PLC models. These include ON/OFF control, proportional (P) control, proportional-integral (PI) control, and proportional-integral-derivative (PID) control. Each involves the use of some form of *closed-loop control* to maintain a process characteristic such as a temperature, pressure, flow, or level at a desired value.

A typical block diagram of a closed-loop control system is shown in Fig. 10-24 (see page 144). A measurement is made of the variable to be controlled. This measurement is then compared to a reference point, or setpoint. If a difference (error) exists between the actual and desired levels, the PLC control program will take the necessary corrective action.

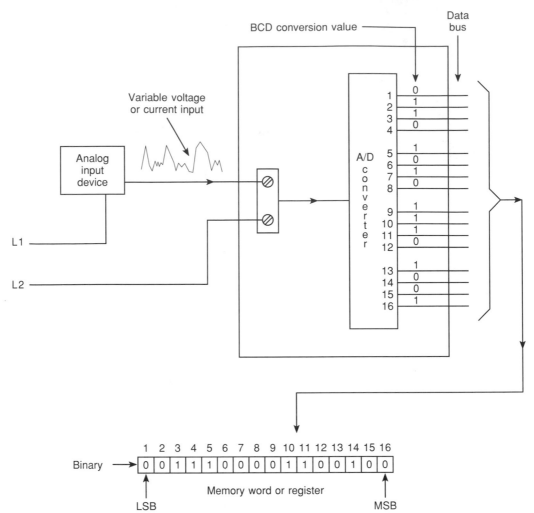

Fig. 10-21 Analog input interface module.

A four BCD thumbwheel switch. *(Courtesy of Cincinnati Milacron)*

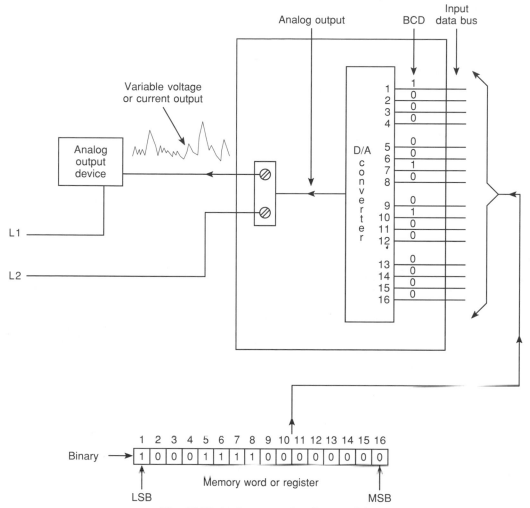

Fig. 10-22 Analog output interface module.

BCD to seven segment display. *(Courtesy of Cincinnati Milacron)*

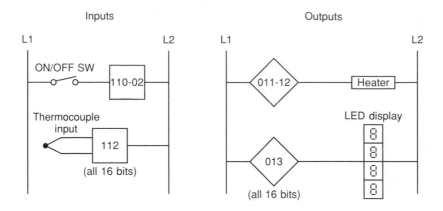

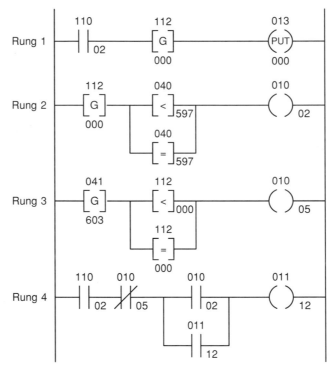

Fig. 10-23 Setpoint temperature control program.

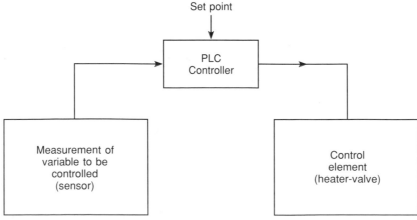

Fig. 10-24 Block diagram of closed-loop control system.

REVIEW QUESTIONS

1. Explain the difference between a register or word and a table or file.

2. What do data manipulation instructions allow the PLC to do?

3. Into what two broad categories can data manipulation instructions be placed?

4. What is involved in a data transfer instruction?

5. Explain what the following logic rung is telling the processor to do (Fig. 10-25):

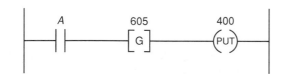

Fig. 10-25

6. What is involved in a data compare instruction?

7. Name and draw the symbol for five different types of data compare instructions.

8. In what way are multibit I/O interfaces different from the discrete type?

9. Assume a thumbwheel switch is set for the decimal number 3286.
   (a) What is the equivalent BCD value for this setting?
   (b) What is the equivalent binary value for this setting?

10. Assume a thermocouple is connected to an analog input module. Explain how the temperature of the thermocouple is communicated to the processor.

11. Name two typical analog output field devices.

12. With the aid of a block diagram, explain the basic operation of a closed-loop control system.

13. Explain what each of the following logic rungs is telling the processor to do (Fig. 10-26):

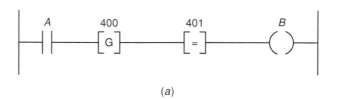

(a)

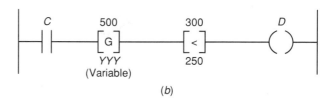

(b)

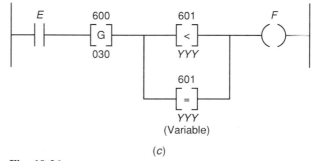

(c)

Fig. 10-26

PROBLEMS

1. Study the data transfer program (see Fig. 10-27) and answer the following questions.
   (a) When S1 is off, what number will appear in the XXX position below the PUT instruction?
   (b) When S1 is on, what number will appear in the XXX position below the PUT instruction?
   (c) When S1 is on, what number will appear in the LED display?
   (d) What is required in order for the number 216 to appear in the LED display?

2. Study the data transfer counter program (see Fig. 10-28) and answer the following questions.
   (a) Where does the three-digit number XXX below the GET instruction in rung 2 come from?

   (b) Outline the steps to follow to operate the program so that output 010/01 is energized after 25 OFF-to-ON transitions of input 110/17.

3. Construct a nonretentive timer program that will turn on a pilot light after a time-delay period. Use a thumbwheel switch to vary the preset time-delay value of the timer.

4. Study the data compare program (see Fig. 10-29) and answer the following questions.
   (a) Will the pilot light come on whenever switch S1 is closed?
   (b) Must switch S1 be closed in order to change the number XXX below the GET instruction?
   (c) What number (XXX) must be below the GET instruction in order to have the pilot light turn on?
   (d) What is word address 055?

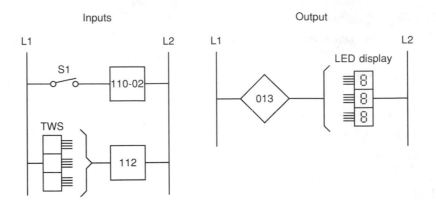

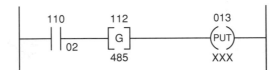

Fig. 10-27

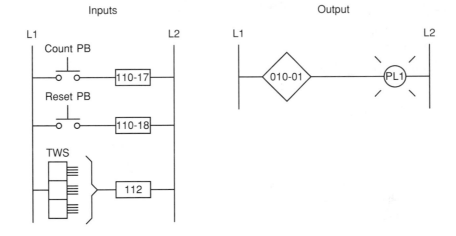

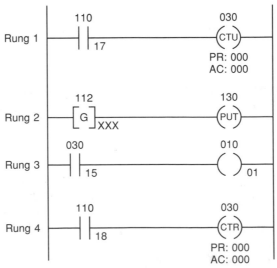

Fig. 10-28

Inputs   Output

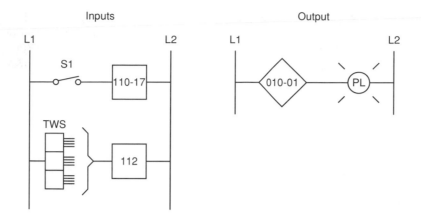

Logic ladder program

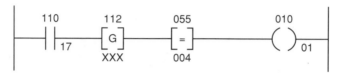

Fig. 10-29

5. Study the data compare program below and answer the following questions (Fig. 10-30):
   (a) List the values that could be found below the GET instruction that would allow the pilot light to turn on.
   (b) If the value in the word 112 is 003 and switch S1 is open, will the pilot light turn on?

(c) Since word 061 is a storage word, could you enter a number other than 012 into the three-digit position below the [<] instruction?

6. Write a program to perform the following:
   (a) Turn on pilot light 1 (PL1) if the thumbwheel switch value is less than 4.

Inputs   Output

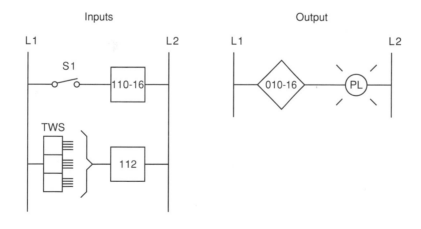

Logic ladder program

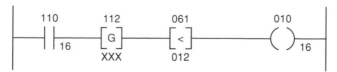

Fig. 10-30

(b) Turn on pilot light 2 (PL2) if the thumbwheel switch value is equal to 4.

(c) Turn on pilot light 3 (PL3) if the thumbwheel switch value is greater than 4.

(d) Turn on pilot light 4 (PL4) if the thumbwheel switch value is less than or equal to 4.

(e) Turn on pilot light 5 (PL5) if the thumbwheel switch value is greater than or equal to 4.

11

MATH INSTRUCTIONS

Upon completion of this chapter you will be able to:

- Analyze and interpret math instructions as they apply to a PLC program
- Create PLC programs involving math instructions
- Analyze and interpret logic ladder programs

11-1 MATH INSTRUCTIONS

Math instructions, like data manipulation instructions, enable the programmable controller to take on some of the qualities of a computer system. The PLC's math functions capability is not intended to allow it to replace a calculator, but rather to allow it to perform arithmetic functions on values stored in memory words.

Depending on what type of processor is being used, various math instructions can be programmed. We will deal only with the four basic math functions: addition (+), subtraction (−), multiplication (×), and division (÷). These instructions use the contents of two words or registers and perform the desired function. The PLC instructions for data manipulation (data transfer and data compare) are used with the math symbols to perform math functions.

11-2 ADDITION INSTRUCTION

The ADD instruction −(+)− performs the additon of two values stored in the referenced memory locations. How these values are accessed is dependent on the con-

troller. Figure 11-1 shows the format used by the Allen-Bradley PLC-2 family of PLCs. Again, even though the format and instructions vary with each manufacturer, the concepts remain the same. In this controller, addition is accomplished by reporting the values stored in two GET instructions immediately followed by the addition instruction. When input device 111/05 is TRUE, the value of word 030 (105) is added to the value of word 031 (080) and the sum (185) is stored in word 032.

The program of Fig. 11-2 shows how the ADD instruction can be used to add the accumulated counts of two up-counters. This application requires a light to come on when the sum of the counts from the two counters is equal to or greater than 350. GET instruction 050 is addressed to get the accumulated value from counter 050. GET instruction 051 is addressed to get the accumulated value from counter 051. These two values are added together by the addition address 020 of rung 3. Rung 4 is a GET/GREATER THAN OR EQUAL TO logic rung. The output in rung 4 will be logic TRUE whenever the accumulated values in the two counters are equal to or greater than the referenced value, 350. A RESET button is provided to reset the accumulated count of both counters to zero.

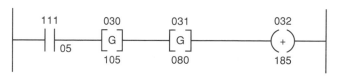

$$(105 + 080 = 185)$$

Fig. 11-1 ADD instruction.

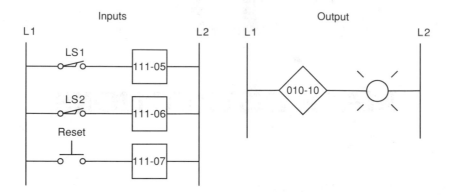

Logic ladder program

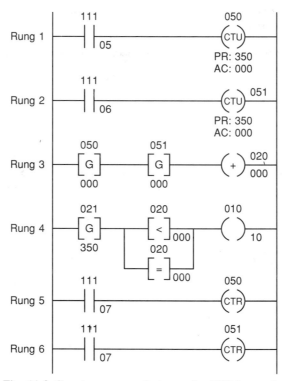

Fig. 11-2 Counter program that uses the ADD instruction.

11-3 SUBTRACTION INSTRUCTION

The SUBTRACT -(−)- instruction is similar to the ADD instruction. However, it uses a store word or register to reflect the difference between, rather than the sum of, two GET values. Figure 11-3 shows a typical SUBTRACT instruction rung. When input device 111/05 is TRUE, the value of word 031 (080) is subtracted from the value of word 030 (105), and the difference (025) is stored in word 032. Only positive values can be used

(105 − 080 = 025)

Fig. 11-3 SUBTRACT instruction.

with some PLCs. When the difference is a negative number and it is used for a subsequent operation, inaccurate results may occur.

The program of Fig. 11-4 shows how the SUBTRACT function can be used to indicate a vessel overfill condition. This application requires an alarm to sound when a supply system leaks 5 lb or more of raw material into the vessel after a preset weight of 500 lb has been reached. When the START button is pressed, the fill solenoid (rung 1) and filling indicating light (rung 2) are turned on and

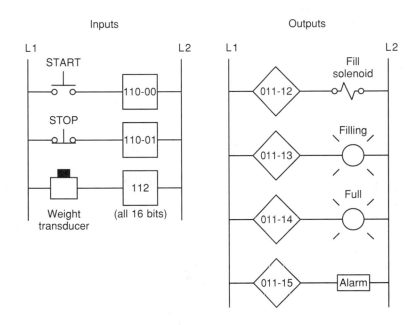

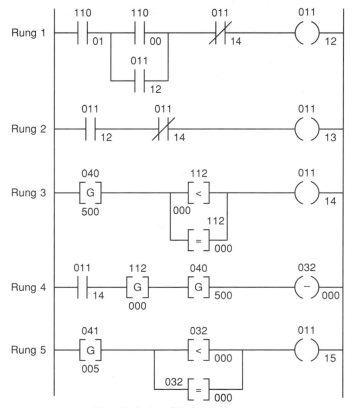

Fig. 11-4 Overfill alarm program.

raw material is allowed to flow into the vessel. The vessel has its weight continuously monitored by the PLC program (rung 3) as it fills. When the weight reaches 500 lb, the fill solenoid is de-energized and the flow is cut off. At the same time, the filling pilot light indicator is turned off and the full pilot light indicator (rung 3) is turned on. Should the fill solenoid leak 5 lb or more of raw material into the vessel, the alarm (rung 5) will energize and stay energized until the overflow level is reduced below the 5-lb overflow limit.

11-4 MULTIPLICATION INSTRUCTION

The MULTIPLY -(×)-(×)- instruction operates on *two* stored memory words instead of one. Two memory words are used so that a number larger than 999 can be displayed and stored. The first word contains the most significant digit (MSD) and the second word contains the least significant digit (LSD). If the product is less than six digits, leading zeros appear in the product. For good documentation habits, the manufacturer recommends using *consecutive* word addresses for the two addresses of the MULTIPLY instruction.

Figure 11-5 shows a simple MULTIPLY program. When input device 111/05 is TRUE, the value of word 030 (123) is multiplied by the value of word 031 (061), and the product (7 503) is stored in word 032 (007) and word 033 (503). As a result rung 2 will become TRUE, turning output 010/10 on.

Figure 11-6 shows how the MULTIPLY function is used as part of an oven temperature control program. In this program the PLC calculates the upper and lower *deadband* or OFF/ON limits about the setpoint. The upper and lower limits are automatically set at ±1 percent regardless of the setpoint value. The setpoint temperature is adjustable by means of thumbwheel switches, and an analog thermocouple interface module is used to monitor the current temperature of the oven. In this example the setpoint temperature is 400°F: therefore, the electric heaters will be turned on when the temperature of the

oven drops to less than 396°F and stay on until the temperature rises above 404°F. If the setpoint is changed to 100°F, the deadband remains at ±1 percent, with the lower limit being 99°F and the upper limit being 101°F. The number stored in word 037 represents the upper temperature limit, while the number stored in word 038 represents the lower temperature limit.

11-5 DIVISION INSTRUCTION

Operation of the DIVIDE -(÷)-(÷)- instruction is very similar to the operation of the MULTIPLY instruction. DIVIDE takes the number stored in the first GET instruction of the division rung and divides it by the number stored in the second GET instruction. The result of the division is held in two words or registers as referenced by the output coils. The first output holds the *integer,* while the second output holds the *decimal fraction.*

Figure 11-7 shows a simple DIVIDE instruction program. When input device 111/05 is TRUE, the value of word 030 (150) is divided by the value of word 031 (040), and the quotient is stored in words 032 and 033 (003.750 or 3.75). As a result rung 2 will become TRUE, turning output 010/10 on.

Figure 11-8 shows how the DIVIDE function is used as part of a program to convert Celsius temperature to Fahrenheit. In this application the thumbwheel switch connected to the input module indicates Celsius temperature. The program is designed to convert the recorded Celsius temperature in the data table to Fahrenheit values for display. The formula

$$°F = \left(\frac{9}{5} \cdot °C \right) + 32$$

forms the basis for the program. In this example, a current temperature reading of 60°C is assumed. In rung 1, the GET instruction at address 030 multiplies the temperature (60°C) by 9 and stores the product (540) in address 032. In rung 2, the GET instruction at address 033

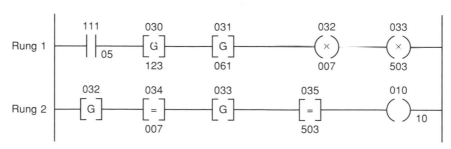

(123 × 061 = 007 503)

Fig. 11-5 MULTIPLY instruction.

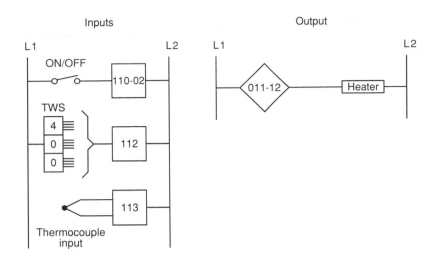

Logic ladder program

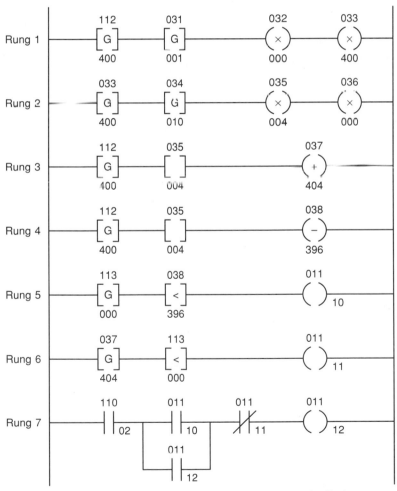

Fig. 11-6 Automatic control of upper and lower setpoint limits.

divides 5 into 540 and stores the quotient 108 in address 034. In rung 3, the GET instruction at address 035 adds 32 to the value of 108 and stores the sum 140 in address 036. Thus 60°C = 140°F. In rung 4, the GET/PUT instruction pair transfers the converted temperature reading, 140°F, to the LED display.

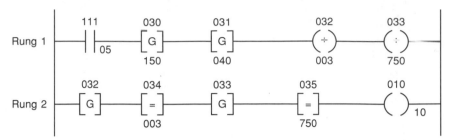

Fig. 11-7 DIVIDE instruction.

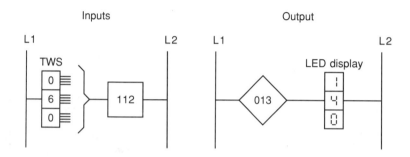

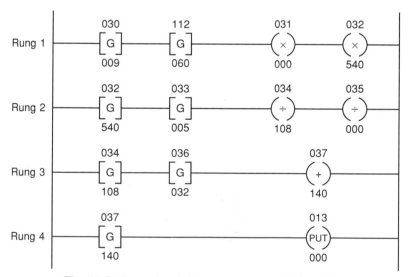

Fig. 11-8 Converting Celsius temperature to Fahrenheit.

REVIEW QUESTIONS

1. Explain the purpose of math instructions as applied to the PLC.

2. What are the four basic math functions that can be performed on some PLCs?

3. What standard format is used for PLC math instructions?

4. Explain what each of the logic rungs is telling the processor to do (see Fig. 11-9):

5. With reference to Fig. 11-10:

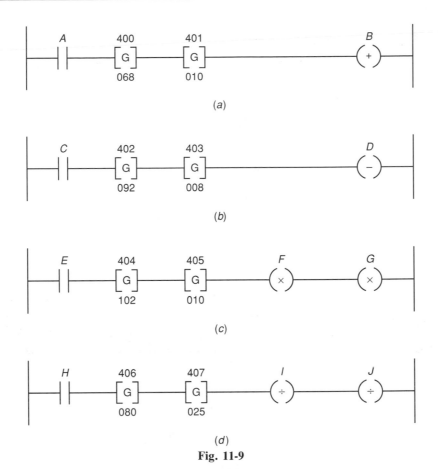

(a)

(b)

(c)

(d)

Fig. 11-9

(a) What is the value of the number stored in word 402?

(b) When will output C be energized?

6. With reference to Fig. 11-11:
   (a) What is the value of the number stored in word 032?
   (b) When will output D be energized?

7. With reference to Fig. 11-12:
   (a) What is the value of the number stored in word 402?

(b) What is the value of the number stored in word 403?

(c) In order for rung 2 to have logic continuity, what numbers must be stored in addresses 404 and 405?

8. With reference to Fig. 11-13:
   (a) What is the value of the number stored in word 402?
   (b) What is the value of the number stored in word 403?
   (c) In order for rung 2 to have logic continuity, what numbers must be stored in words 404 and 405?

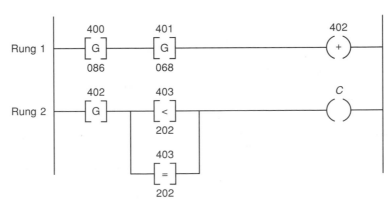

Fig. 11-10

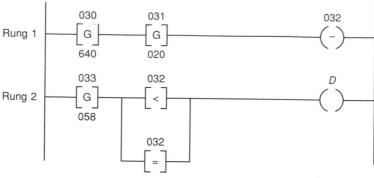

Fig. 11-11

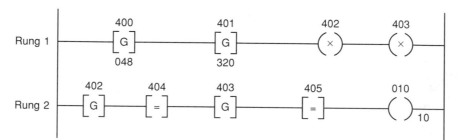

Fig. 11-12

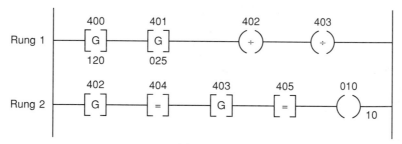

Fig. 11-13

PROBLEMS

1. Answer each of the following with reference to the *counter program* shown in Fig 11-2.
   (a) Assume the accumulated count of counters 050 and 051 to be 148 and 036 respectively. State the value of the number stored in each of the following words at this point: (i) 050, (ii) 051, (iii) 020, (iv) 021.
   (b) Will output 010/10 be energized at this point? Why?
   (c) Assume the accumulated count of counters 050 and 051 to be 250 and 175 respectively. State the value of the number stored in each of the following words at this point: (i) 050, (ii) 051, (iii) 020, (iv) 021.
   (d) Will output 010/10 be energized at this point? Why?

2. Answer each of the following with reference to the *overfill alarm program* shown in Fig. 11-4.
   (a) Assume that the vessel is filling and has reached the 300-lb point. State the status of each of the logic rungs (TRUE or FALSE) at this point.
   (b) Assume that the vessel is filling and has reached the 480-lb point. State the value of the number stored in each of the following words at this point: (i) 040, (ii) 112, (iii) 032, (iv) 041.
   (c) Assume that the vessel is filled to a weight of 502 lb. State the status of each of the logic rungs (TRUE or FALSE) for this condition.
   (d) Assume that the vessel is filled to a weight of 510 lb. State the value of the number stored in each of the following words for this condition: (i) 040, (ii) 112, (iii) 032, (iv) 041.
   (e) With the vessel filled to a weight of 510 lb, state the status of each of the logic rungs (TRUE or FALSE).

3. Answer the following with reference to the *upper and lower temperature control program* shown in Fig 11-6.
   (a) Assuming that the setpoint temperature is 600°F, at what temperature will the electric heaters be turned on and off?
   (b) Assuming that the setpoint temperature is 600°F and the thermocouple input module indicates a temperature of 590°F, what is the value of the number stored in each of the following words at this point: (i) 112, (ii) 031, (iii) 032, (iv) 033, (v) 034, (vi) 035, (vii) 036, (viii) 037, (ix) 038, (x) 113?

   (c) Assuming that the setpoint temperature is 600°F and the thermocouple input module indicates a temperature of 608°F, what is the status (energized or not energized) of each of the following outputs: (i) 011/10, (ii) 011/11, (iii) 011/12?

4. With reference to the *converting Celsius temperature to Fahrenheit program* shown in Fig. 11-8 state the value of the number stored in each of the following words for a thumbwheel setting of 035: (i) 030, (ii) 112, (iii) 031, (iv) 032, (v) 033, (vi) 034, (vii) 035, (viii) 036, (ix) 037, (x) 013.

12

SEQUENCER INSTRUCTIONS

Upon completion of this chapter you will be able to:

- Identify and describe the various forms of mechanical sequencers and explain the basic operation of each
- Interpret and explain information associated with a PLC sequence instruction
- Compare the operation of an event-driven and a time-driven sequencer

12-1 MECHANICAL SEQUENCERS

Sequencer instructions are named after the mechanical sequencer switches they replace. These switches are often referred to as *drum switches, rotary switches, step-* *per switches,* or *cam switches,* in addition to the *sequencer switch* identification. Figure 12-1 illustrates the operation of a cam-operated sequencer switch. An electric motor is used to drive the cams. A series of leaf-spring–mounted contacts interacts with the cam so that

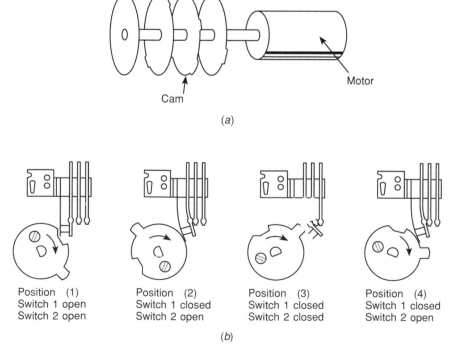

(a)

Position (1)
Switch 1 open
Switch 2 open

Position (2)
Switch 1 closed
Switch 2 open

Position (3)
Switch 1 closed
Switch 2 closed

Position (4)
Switch 1 closed
Switch 2 open

(b)

Fig. 12-1 Mechanical cam-operated sequencer. (*a*) Cam-driving mechanism. (*b*) Cam and contact operation.

Typical industrial rotating cam limit switch is a pilot circuit device used with machinery having a repetitive cycle of operation. *(Courtesy of Allen-Bradley Company, Inc.)*

in different degrees of rotation of the cam various contacts are closed and opened to energize and de-energize various electrical devices.

Figure 12-2 illustrates a typical mechanical drum-operated sequencer switch. The switch consists of series of contacts that are operated by pegs located on a motor-driven drum. The pegs can be placed at random locations around the circumference of the drum to operate contacts. When the drum is rotated, contacts that align with the pegs will close, while the contacts where there are no pegs will remain open. In this example the pres-

nce of a peg could be thought of as logic 1, or ON, while the absence of a peg would be logic 0, or OFF.

Figure 12-2 also shows an equivalent sequencer data table for the drum cylinder. If the first five steps on the drum cylinder were removed and flattened out, they would appear as illustrated in the table. Each horizontal location where there was a peg is now represented by a 1 (ON), and the positions where there were no pegs are each represented by a 0 (OFF).

Sequencer switches are used whenever a repeatable operating pattern is required. An excellent example is the sequencer switch used in dishwashers to pilot the machinery through a wash cycle (Fig. 12-3). The cycle is always the same, and each step occurs for a specific time. The domestic washing machine is another example of the use of a sequencer, as are dryers and similar time-clock–controlled devices.

Figure 12-4 shows the wiring diagram and data table for a dishwasher that uses a cam-operated sequencer commonly known as the *timer*. In this unit a synchronous motor drives a mechanical train, which, in turn, drives a series of cam wheels. The cam advances in increments of about 45 s in duration. Normally, the timer motor operates continuously throughout the cycle of operation. The data table in Fig. 12-4 outlines the sequence of operation of the timer. Each time increment is 45 s. A total of sixty 45-s steps are used to complete the 45-min operating cycle. The numbers in the "Active Circuits" column refer to the circled numbers found on the schematic diagram.

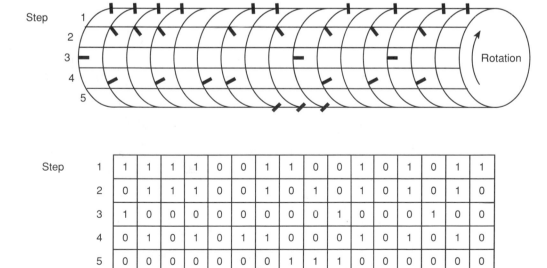

Peg locations in cylinder

Step																
1	1	1	1	1	0	0	1	1	0	0	1	0	1	0	1	1
2	0	1	1	1	0	0	1	0	1	0	1	0	1	0	1	0
3	1	0	0	0	0	0	0	0	0	1	0	0	0	1	0	0
4	0	1	0	1	0	1	1	0	0	0	1	0	1	0	1	0
5	0	0	0	0	0	0	0	1	1	1	0	0	0	0	0	0

Equivalent sequencer
data table

Fig. 12-2 Mechanical drum-operated sequencer.

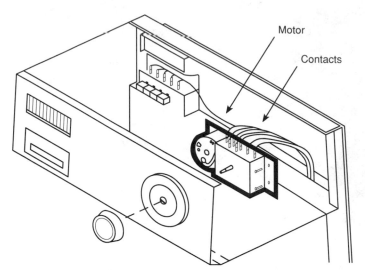

Fig. 12-3 Dishwasher sequence.

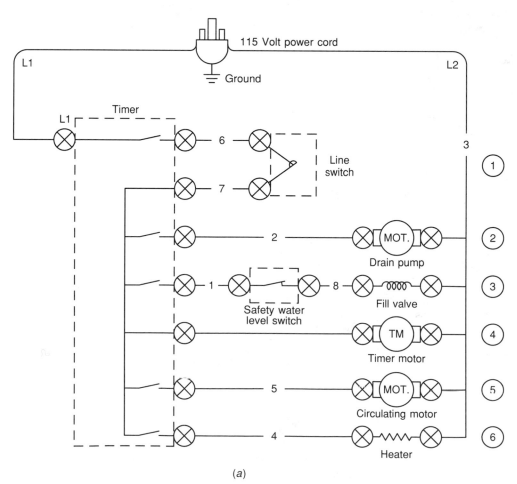

(a)

Fig. 12-4 Dishwasher wiring diagram and data table.

Machine function		Timer increment	Active circuits					
OFF		0-1						
1st Prerinse	Drain	2	1	2		4		
	Fill	3	1		3	4	5	
	Rinse	4-5	1			4	5	6
	Drain	6	1	2		4	5	
Prewash	Fill	7	1		3	4	5	
	Wash	8-10	1			4	5	6
	Drain	11	1	2		4	5	
2nd Prerinse	Fill	12	1		3	4	5	
	Rinse	13-15	1			4	5	6
	Drain	16	1	2		4		
Wash	Fill	17	1		3	4		
	Wash	18-30	1			4	5	6
	Drain	31	1	2		4	5	
1st Rinse	Fill	32	1		3	4	5	
	Rinse	33-34	1			4	5	6
	Drain	35	1	2		4	5	
2nd Rinse	Fill	36	1		3	4	5	
	Rinse	37-41	1			4	5	6
	Drain	42	1	2		4	5	
Dry	Dry	43-58	1			4		6
	Drain	59	1	2		4		6
	Dry	60	1			4		6

(b)

Fig. 12-4 (*Cont.*)

12-2 SEQUENCER INSTRUCTIONS

The sequencer instruction is a powerful instruction found on many PLCs today. A programmed sequencer can replace a mechanical drum switch. To program a sequencer, binary information is entered into a series of consecutive memory words. These consecutive memory words are referred to as a *word file*. Data is first entered into a word file for each sequencer step. As the sequencer advances through the steps, binary information is transferred from the word file to the output word(s).

Figure 12-5 illustrates how the sequencer output instruction works. In this example, 16 lights are used for outputs. Each light represents one bit address (1 through 16) of output word 050. The lights are programmed in a four-step sequence to simulate the operation of two-way traffic lights.

Data is entered into a word file for each sequencer step as illustrated in Fig. 12-6. In this example, words 60, 61, 62, and 63 are used for the four-word file. Using

the programmer, binary information (1's and 0's) that reflects the desired light sequence is entered into each word of the file. For ease of programming, some PLCs allow the word file data to be entered using the octal, hexadecimal, BCD, or similar number system. When this is the case, the required binary information for each sequencer step must first be converted to whatever number system is employed by the PLC. This information is then entered with the programmer into the word file.

Once the data has been entered into the word file of the sequencer, the PLC is ready to control the lights. When the sequencer is activated and advanced to *step 1*, the binary information in word 060 of the file is transferred into word 050 of the output. As a result, lights 1 and 12 will be switched on and all the rest will remain off. Advancing the sequencer to *step 2* will transfer the data from word 061 into word 050. As a result, lights 1, 8, and 12 will be on and all the rest will be off. Advancing the sequencer to *step 3* will transfer the data from word 062 into word 050. As a result, lights 4 and 9 will

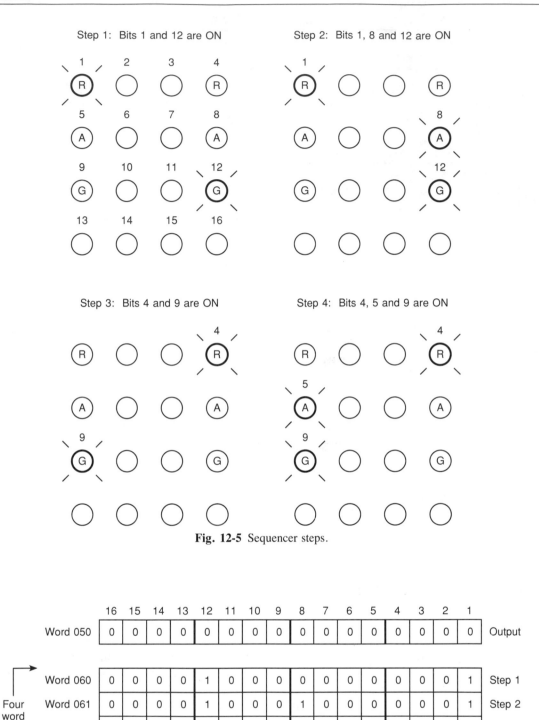

Fig. 12-5 Sequencer steps.

Fig. 12-6 Binary information for each sequencer step.

be on and all the rest will be off. Advancing the sequencer to *step 4* will transfer the data from word 063 into word 050. As a result, lights 4, 5, and 9 will be on and all the rest will be off. When the last step is reached, the sequencer is either automatically or manually reset to step 1.

When a sequencer operates on an entire output word, there may be outputs associated with the word that do

not need to be controlled by the sequencer. In our example, bits 2, 3, 5, 7, 10, 11, 13, 14, 15, and 16 of output word 050 are not used by the sequencer but could be used elsewhere in your program.

To prevent the sequencer from controlling these bits of the output word, a *mask* word (040) is used. The use of a mask word is illustrated in Fig. 12-7. The mask

	16	15	14	13	12	11	10	9	8	7	6	5	4	3	2	1	
Word 050	0	0	0	0	0	0	0	0	0	0	0	0	0	0	0	0	Output
Word 040	0	0	0	0	1	0	0	1	1	0	0	1	1	0	0	1	Mask
Word 060	0	0	0	0	1	0	0	0	0	0	0	0	0	0	0	1	Step 1
Word 061	0	0	0	0	1	0	0	0	1	0	0	0	0	0	0	1	Step 2
Word 062	0	0	0	0	0	0	0	1	0	0	0	0	1	0	0	0	Step 3
Word 063	0	0	0	0	0	0	0	1	0	0	0	1	1	0	0	0	Step 4

Fig. 12-7 Using a mask word.

word selectively screens out data from the sequencer word file to the output word. For each bit of output word 050 that the sequencer is to control, the corresponding bit of mask word 040 must be set to 1. All other bits of output word 050 are set to 0 and so can be used independently of the sequencer.

Sequencers, like other PLC instructions, are programmed differently with each PLC, but again the concepts are the same. The advantage of sequencer programming over the conventional program is the large savings of memory words. Typically, the sequencer program can do in 20 words or less what a standard program can do in 100 words.

Sequencer output instructions can be either block- or coil-formatted and contain a counter or timer and a file. The instructions require the entry of more than one address. A typical PLC block-formatted sequencer instruction is illustrated in Fig. 12-8. Sequencer instructions are usually retentive and there is an upper limit as to the number of external outputs and steps that can be operated upon by a single instruction. Many sequencer instructions reset the sequencer automatically to step 1 upon completion of the last sequence step. Other instructions provide an individual reset control line or a combination of both.

12-3 SEQUENCER PROGRAM

A sequencer program can be *event-driven* or *time-driven*. An event-driven sequencer operates similarly to a mechanical stepper switch that increments by one step for each pulse applied to it. A time-driven sequencer operates similarly to a mechanical drum switch that increments automatically after a preset time period.

Figure 12-9 shows a typical sequencer data worksheet that is used to document the different sequencer steps before they are programmed into the PLC. This form allows you to document sequencer data in an orderly, systematic way, reducing the chances for programming errors.

As mentioned, sequencer numerical data is often programmed using a shorter notation than the basic binary number system. In the worksheet example of Fig. 12-9, sequencer mask and output status data must be converted from binary to hexadecimal for programming purposes. The table of Fig. 12-10 is used to aid the programmer in making this conversion.

The sequencer data worksheet also calls for the identification of the I/O module group numbers and corresponding terminal addresses. Figure 12-11 shows a typical identification chart for these locations. In this particular

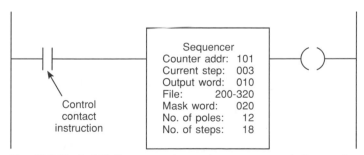

Fig. 12-8 Typical PLC sequencer instruction. *Courtesy of Allen-Bradley Company, Inc.)*

SEQUENCER INSTRUCTION DATA FORM

SEQUENCER CLASSIFICATION: —(SQI)— ☐   —(SQO)— ☐   ADDRESS **9** : :   TIME DRIVEN ☐   EVENT DRIVEN ☐

	SEQUENCER DATA										PROGRAM ENTRY CODE		
EVENT DESCRIPTION	SEQUENCER DATA GROUP →	SECOND			FIRST						2nd Grp	1st Grp	PRESET
	MODULE GROUP NUMBER →												
	I/O TERMINAL ADDRESS →												
	MASK DATA →												
	STEP NO. 1												
	2												
	3												
	4												
	5												
	6												
	7												
	8												
	9												
	10												
	11												
	12												
	13												
	14												
	15												
	16												
	17												
	18												
	19												
	20												

Fig. 12-9 Typical sequencer instruction data form.   *Courtesy of Allen-Bradley Company, Inc.)*

Sequencer data	Program code
0000	0
0001	1
0010	2
0011	3
0100	4
0101	5
0110	6
0111	7
1000	8
1001	9
1010	A
1011	B
1100	C
1101	D
1110	E
1111	F

Example - For sequencer data 0001 1111, enter code F for 1st Sequencer Data Group, and code 1 for 2nd Sequencer Data Group. Thus: 0001 1111 = 1 F

Fig. 12-10 Converting from binary to hexadecimal.

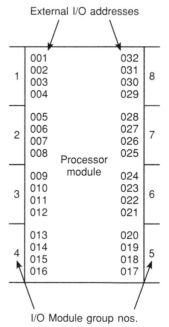

Fig. 12-11 I/O module identification chart.

controller, you have the option of assigning any of the I/O module group numbers to either the first or second sequencer data group. Under each I/O module group number you must list the corresponding four terminal addresses in reverse order, beginning with the highest number at the left.

Figure 12-12 shows the program for a coil-formatted, time-driven sequencer. This four-step, time-driven se-

quencer controls six external outputs. The status of the outputs, for each step, are as recorded in the data worksheet form. Outputs 015 and 016 are masked so they can be used elsewhere in the program. The program entry code data is simply the mask and output status data converted from binary to hexadecimal code.

With time-driven sequencers, each step has a function similar to a timer instruction, in that it involves an ac

Sequencer instruction data form

Address ___SQO 901___   Time driven [X]   Event driven [ ]

Event description	Mask data →	Second				First				2nd Grp	1st Grp	Preset (Time)
	Module group number →	4				3						
	I/O Terminal address →	016	015	014	013	012	011	010	009			
	Mask data →	0	0	1	1	1	1	1	1	3	F	
Red – Green	Step no. 1	0	0	1	0	0	0	0	1	2	1	30 s
Red – Green & Amber	2	0	0	1	0	1	0	0	1	2	9	5 s
Green – Red	3	0	0	0	1	0	0	1	0	1	2	30 s
Green & Amber – Red	4	0	0	0	1	0	1	1	0	1	6	5 s

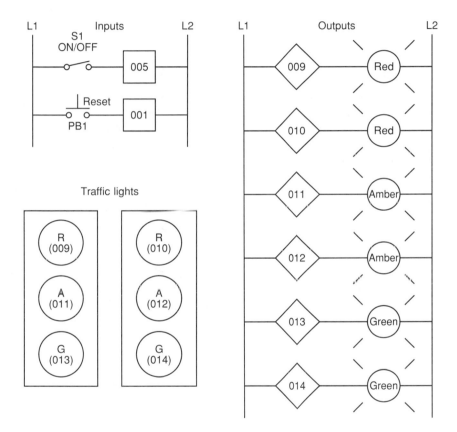

Traffic lights

Logic ladder program

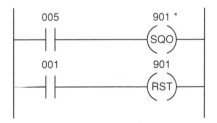

* Information from the sequencer data form worksheet is entered into the sequencer 901 output file instruction. Once the SQO instruction is entered the display will prompt you for the required data.

Fig. 12-12 Time-driven sequencer.

Sequencer instruction data form

Address **SQO 901**   Time driven ☐   Event driven ☒

			Second				First				Program entry code		Preset (Count)
Sequencer data group →			Second				First						
Module group number →			4				3						
I/O Terminal address →		016	015	014	013	012	011	010	009	2nd Grp	1st Grp	
Event description	Mask data →	0	0	1	1	1	1	1	1	3	F	
Red – Green	Step no. 1	0	0	1	0	0	0	0	1	2	1	1
Red – Green & Amber	2	0	0	1	0	1	0	0	1	2	9	1
Green – Red	3	0	0	0	1	0	0	1	0	1	2	1
Green & Amber – Red	4	0	0	0	1	0	1	1	0	1	6	1

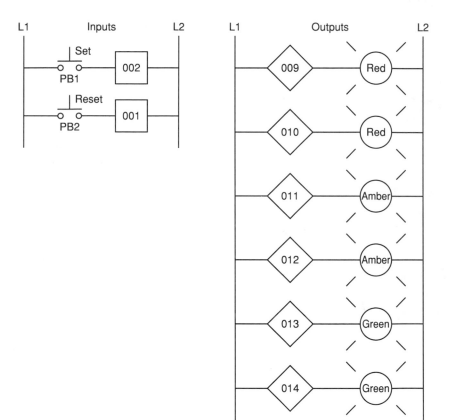

Logic ladder program

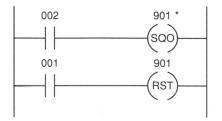

* Information from the sequencer data worksheet is entered into the sequencer 901 output file instruction. Preset time periods are replaced with preset counts and sequencer designation is changed from time-driven to event-driven.

Fig. 12-13 Event-driven sequencer.

cumulated time value and a programmed, preset time value. The preset time values for each step are as re corded in the data worksheet form.

This particular program (Fig. 12-12) simulates the operation of two-way traffic lights and is similar to the light sequence of Fig. 12-5. Beginning with the sequencer reset, when the SQ0 instruction goes TRUE, step 1 is initiated. As a result, outputs 009 and 014 are on and all the rest are off. After a preset time of 30 s (assuming SQ0 remains TRUE), step 2 is initiated. As a result, outputs 009, 012, and 014 are on and all the rest are off. After 5 s, step 3 begins and outputs 010 and 013 are on and all the rest are off. After 30 s, step 4 begins and outputs 010, 011, and 013 are on and all the rest are off. After 5 s, the cycle automatically repeats with step 1. The sequence can be stopped at any step by opening the ON/OFF switch S1. The sequence can be reset to step 1 at any time by pressing RESET button PB1.

In Fig. 12-13 you can see the traffic light circuit of Fig. 12-12 programmed as an event-driven sequencer. The event-driven sequencer functions similarly to the counter instruction, in that it involves an accumulated counter value and a programmed, preset counter value. If the preset counter value is set to 1, closing and opening step push button PB1 will manually step you through the different sequencer steps. Beginning with the sequencer reset, step 1 is initiated and remains so until step button PB1 is actuated. Momentarily closing PB1 produces a FALSE-to-TRUE transition that increments the sequencer to step 2. Thus the event-driven sequencer counts FALSE-to-TRUE transitions of the sequencer rung. When the accumulated count value reaches the preset count value, the sequencer advances to the next step and the accumulated value increments from zero again. If the preset count value were set for 2, it would take two FALSE-to-TRUE transitions of the sequencer rung to advance the sequence one step.

REVIEW QUESTIONS

1. Explain the basic operation of a cam-operated sequencer switch.

2. What type of operations are sequencer switches most suitable for?

3. What is the advantage of sequencer programming over conventional programming methods?

4. With reference to the PLC sequencer instruction:
   (a) Where is the information for each sequencer step entered?
   (b) What is the function of the output word of a sequencer instruction?
   (c) Explain the transfer of data that occurs as the sequencer is advanced through its various steps.

5. Explain the purpose of a mask word when used in conjunction with the sequencer instruction.

6. What are the two limits placed on sequencer instructions?

7. Sequencer instructions are usually retentive. Explain what this means.

8. Explain the difference between an event-driven and a time-driven sequencer.

PROBLEMS

1. Answer each of the following with reference to the *dishwasher circuit* shown in Fig. 12-4.
   (a) How many cam switches can be found in the timer?
   (b) How many timed steps are there for one complete operating cycle?
   (c) What is the value of the time interval for each step?
   (d) State the five output devices operated by the timer.
   (e) What is the total length of time that the heater is on for one complete cycle?
   (f) What output devices would normally be on when the timer is at the 20-min point in the cycle?
   (g) What is the greatest length of time that the fill valve stays energized?
   (h) Explain the function of the safety water level switch.
   (i) Outline the sequence in which the outputs are energized for the first RINSE portion of the cycle.
   (j) Why is the timer motor off for only one step in the entire operating cycle?

2. Construct an equivalent sequencer data table for the six steps of the drum-operated sequencer drawn in Fig. 12-14:

3. Answer the following with reference to the sequencer word file (Fig. 12-15):
   (a) Assuming output bit addresses 1 through 16 are controlling lights 1 through 16, state the status of each light for each of the steps.

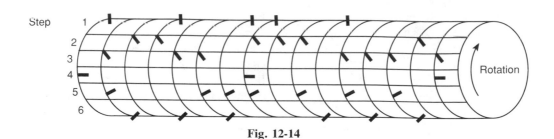

Fig. 12-14

(b) What output bit addresses could be masked?

(c) State the status of each bit of output word 25 for step 3 of the sequencer cycle.

(d) If word 31 is to be entered into the PLC using the hexadecimal code, how would it be written?

(f) What is the total time required for one complete cycle of the sequencer?

(g) State for what step(s) of the sequencer each of the following outputs would be on:

	16	15	14	13	12	11	10	9	8	7	6	5	4	3	2	1	
Word 25	0	0	0	0	0	0	0	0	0	0	0	0	0	0	0	0	Output

Word 30	1	1	1	1	1	1	1	1	1	1	1	1	1	1	1	1	Step 1
Word 31	1	0	1	0	1	0	1	0	1	0	1	0	1	0	1	0	Step 2
Word 32	0	1	0	1	0	1	0	1	0	1	0	1	0	1	0	1	Step 3
Word 33	0	0	0	0	0	0	0	0	0	0	0	0	0	0	0	0	Step 4

Fig. 12-15

4. Answer each of the following with reference to the time-driven sequencer traffic light program shown in Fig. 12-12.

(a) How many bit addresses are controlled by this sequencer?

(b) What is the mask data program code for this sequencer?

(c) Assuming normal operation, when input 005 is turned on, the sequencer will perform the steps as they have been programmed. If input 005 is turned off during normal operation, what happens to the sequencer at that point and when input 005 is turned back on again?

(d) Suppose input 005 were turned off at step 3 in the RUN mode. What outputs would be energized?

(e) If you wanted to control outputs 010, 011, 014, 015, and 016, what would the program entry code for your mask data be?

(i) 009	(iv) 012	(vii) 015
(ii) 010	(v) 013	(viii) 016
(iii) 011	(vi) 014	

(h) Pressing the RESET button resets the sequencer to what step?

(i) What outputs, not used by the sequencer, can be used elsewhere in the program?

5. Answer each of the following with reference to the event-driven sequencer traffic light program shown in Fig. 12-13.

(a) Which input condition is made TRUE to reset the sequencer to step 1?

(b) When does the sequencer advance to the next step?

(c) Assume power to the sequencer is lost and then returned. In what way will the operation of the sequencer be affected?

(d) If the preset value of step 3 were changed from 1 to 2, how would this affect the operation of the sequencer?

13

PLC INSTALLATION PRACTICES, EDITING, AND TROUBLESHOOTING

Upon completion of this chapter you will be able to:

- Outline and describe requirements for a PLC enclosure
- Identify and describe the functions of bleeder resistors in PLCs
- Differentiate between off-line and on-line programming
- Describe proper grounding practices and preventive maintenance tasks associated with PLC systems

13-1 PLC ENCLOSURES

A PLC system, if installed properly, should give years of trouble-free service. The design nature of PLCs includes a number of rugged design features that allow them to be installed in almost any industrial environment. However, problems can occur if the system is not installed properly.

Programmable logic controllers are generally placed within an enclosure. An enclosure is the chief protection from atmospheric conditions. The National Electrical Manufacturers Association (NEMA) has defined enclosure types, based on the degree of protection an enclosure will provide. For most solid-state control devices a NEMA 12 enclosure is recommended. This type of enclosure is for general purpose areas and is designed to be dust-tight. In addition, metal enclosures also help to minimize the effects of electromagnetic radiation that may be generated by surrounding equipment.

Every PLC installation will dissipate heat from its power supplies, local I/O racks, and processor. This heat accumulates in the enclosure and must be dissipated from it into the surrounding air. For many applications normal convection cooling will keep the controller components within specified temperature operating range. Proper spacing of components within the enclosure is usually sufficient for heat dissipation. The temperature inside the enclosure must not exceed the maximum operating temperature of the controller (typically 60°C max.) Additional cooling provisions, such as a fan or blower, may be required where high ambient temperatures are encountered. Figure 13-1 shows the typical layout of components for a PLC installation.

13-2 ELECTRICAL NOISE

When the PLC is operated in a noise-polluted industrial environment, special consideration should be given to possible electrical interference. Malfunctions resulting from noise are temporary occurrences of operating errors that can result in hazardous machine operation in certain applications. Noise usually enters through input, output, and power supply lines. Noise may be coupled into these lines by an electrostatic field or through electromagnetic induction (EMI).

To increase the operating noise margin, the controller should be located away from noise-generating devices such as large ac motors and high-frequency welders. Potential noise generators include relays, solenoids, motors, and motor starters, especially when operated by hard contacts such as push buttons or selector switches. Suppression for noise generation may be necessary when these

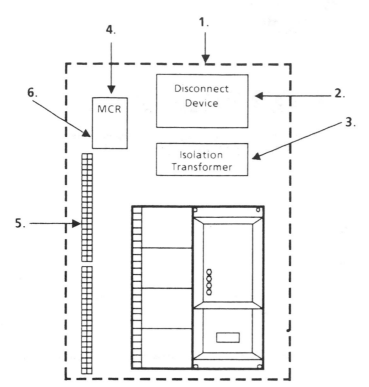

1. A NEMA rated enclosure suitable for your application and environment which will shield your controller from electrical noise and airborne contaminants.

2. A disconnect, to remove power from the system.

3. A fused isolation transformer or a constant voltage transformer, as your application dictates.

4. A master control relay/emergency stop circuit.

5. Terminal blocks or wiring ducts.

6. Suppression devices for limiting EMI generation.

Fig. 13-1 Typical PLC installation. *(Courtesy of Allen-Bradley Company, Inc.)*

PLC system mounted within an enclosure. *(Courtesy of Klockner-Moeller Ltd.)*

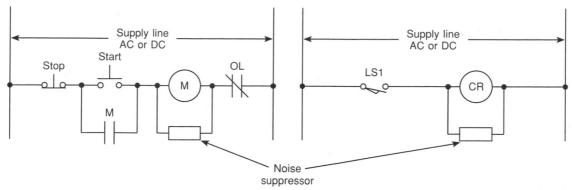

Fig. 13-2 Typical noise suppression methods.

types of loads are connected as output devices. Figure 13-2 shows typical noise suppression methods.

Careful wire routing also helps to cut down on electrical noise. Within the PLC enclosure, input power to the processor module should be routed separately from the wiring to I/O modules. *Never* run signal wiring and power wiring in the same conduit. Segregate I/O wiring by signal type, and bundle wiring with similar electrical characteristics together. Wiring with different signal characteristics should be routed into the enclosure by separate paths whenever possible. A fiber optic system, which is totally immune to all kinds of electrical interference, can also be used for signal wiring.

13-3 LEAKY INPUTS AND OUTPUTS

Many field input devices, such as proximity switches, used with PLC-based systems are of a solid-state design. Any electronically based input sensor that uses a solid-switch silicon controlled rectifier (SCR), triac, or transistor, will have a small leakage current even when in the OFF state. Often, the leaky input will only cause the module's input indicator to flicker. The leakage may, however, result in a falsely activated PLC input. In order to correct the problem, a bleeder resistor is connected across or in parallel with the input as shown in Fig. 13-3.

This leakage may also occur with the solid-state switch used in many output modules. A similar problem can be created when a high-impedance output load device is used with these modules. Figure 13-4 shows how a bleeder resistor is connected to bleed off this unwanted leakage current.

13-4 GROUNDING

Proper grounding is an important safety measure in all electrical installations. The authoritative source on grounding requirements for a PLC installation is the National

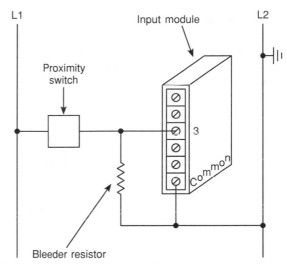

Fig. 13-3 Connection for leaky input devices.

Electrical Code. The code specifies the type of conductors, color codes, and connections necessary for safe grounding of electrical components. According to the code, the grounding path must be permanent (no solder),

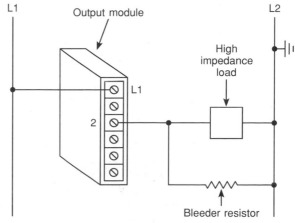

Fig. 13-4 Connection for leaky output devices.

continuous, and able to conduct safely the ground-fault current in the system with minimal impedance. In addition to the grounding required for the controller and its enclosure, you must also provide proper grounding for all controlled devices in your application. Most manufacturers provide detailed information on the proper grounding methods to use in an enclosure. Figure 13-5 shows typical grounding connections for an enclosure.

With solid-state control systems, grounding helps to limit the effects of noise due to electromagnetic induction (EMI). The following grounding practices will help reduce electrical noise interference:

- All PLC equipment and enclosure backplates should be individually grounded to a central point on the enclosure frame.
- Ground wires should be separated from power wiring at the point of entry to the enclosure.
- All ground connections should be made with star washers between the grounding wire and lug and metal enclosure surface.
- Paint or other nonconductive material should be scraped away from the area where a chassis makes contact with the enclosure.
- The minimum ground wire size should be No. 12 AWG stranded copper for PLC equipment grounds and No. 8 AWG stranded copper for enclosure backplate grounds.

- The enclosure should be properly grounded to the ground bus.
- The machine ground should be connected to the enclosure and to earth ground.

13-5 VOLTAGE VARIATIONS AND SURGES

The power supply section of the PLC system is built to sustain line fluctuations and still allow the system to function within its operating range. Where line voltage variation is excessive, a constant voltage transformer can be used to solve the problem. The constant voltage transformer stabilizes the input voltage by compensating for voltage changes at the primary in order to maintain a steady voltage at the secondary.

When current in an inductive load is interrupted or turned off, a very high voltage spike is generated. These voltage spikes, if not suppressed, can reach several thousand volts and produce surges of damaging high currents. To avoid this situation, a suppression network should be installed to limit the voltage spike as well as the rate of change of current through the inductor. Generally, output modules designed to drive inductive loads include suppression networks as part of the module circuit. Figure 13-6 illustrates different methods of suppressing dc and ac inductive loads.

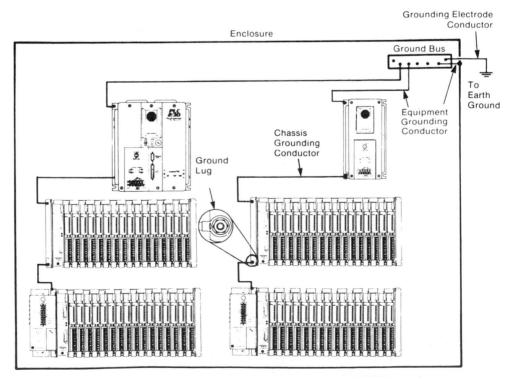

Note: When using this grounding configuration, make no connections to EQUIP GND on the power supply terminal strips. This can cause ground loops.

Fig. 13-5 PLC ground connections. *(Courtesy of Allen-Bradley Company, Inc.)*

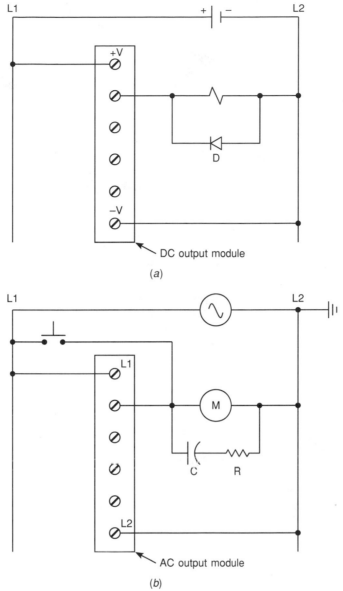

(a)

(b)

Fig. 13-6 Suppressing inductive loads. (a) Suppression of DC load. (b) Suppression of AC load.

Typical surge suppressor devices. *(Courtesy of Allen-Bradley Company, Inc.)*

13-6 PROGRAM EDITING

A feature offered with many PLC programmers is program *editing*. Editing is simply the ability to make changes to an existing program through a variety of editing functions. Using the editing function, instructions and rungs can be added or deleted: addresses, data, and bits can be changed. Again, the editing format varies with different manufacturers and PLC models. Table 13-1 outlines the editing format used with the Allen-Bradley PLC-2/05 processor.

When editing a processor's logic, the use of the SEARCH function can be extremely helpful. This function is used to search the program for specific addressed instructions. Activation of the SEARCH key locates the specified addressed instruction in the processor memory. The circuit containing the searched instruction is then automatically displayed on the screen for user inspection. If desired, the user can then modify the instruction itself, or the circuit containing the instruction. Upon completion of any changes, the SEARCH key can be depressed again for location of the next successive use of the same instruction.

Another important editing technique is the use of the *cursor control* keys. There are usually four keys that are used for cursor control. These keys are usually engraved with arrows pointing up, down, right, and left (Fig. 13-7). Usually a blinking instruction in a ladder rung of the operator terminal display indicates the cursor position. The cursor controls allow you to move through your program from instruction to instruction or from rung to rung.

13-7 PROGRAMMING AND MONITORING

When programming a PLC, several instruction entry modes are available, depending on the manufacturer and the model of the unit. *Off-line* programming requires that the processor be stopped from scanning for the programming operation to occur. When the PLC is placed in the *off-line program mode,* the processor scan automatically stops and all output devices are turned to their OFF state. The off-line program mode is the safest manner in which to edit a program because additions, changes, and deletions do not affect the operation of the system

Table 13-1 TYPICAL EDITING FUNCTIONS

Function	Key Sequence	Mode	Description
Inserting a condition instruction	(INSERT) (Instruction) (Address) or (INSERT) (←) (Instruction) (Address)	Program	Position the cursor on the instruction that will precede the instruction to be inserted. Then press key sequence. Position the cursor on the instruction that will follow the instruction to be inserted. Then press key sequence
Removing a condition instruction	(REMOVE) (Instruction)	Program	Position the cursor on the instruction to be removed and press the key sequence
Inserting a rung	(INSERT) (RUNG)	Program	Position the cursor on any instruction in the preceding rung and press the key sequence. Enter instructions and complete the rung.
Removing a rung	(REMOVE) (RUNG)	Program	Position the cursor anywhere on the rung to be removed and press the key sequence. **IMPORTANT: ONLY ADDRESSES CORRESPONDING TO OUTPUT ENERGIZE, LATCH AND UNLATCH INSTRUCTIONS ARE CLEARED TO ZERO.**
Changing data of a word or block instruction	(INSERT) (Data)	Program	Position the cursor on the word or block instruction whose data is to be changed. Press the key sequence.
Changing the address of a word or block instruction	(INSERT) (First Digit) (←) (Address)	Program	Position the cursor on a word or block instruction with data and press (INSERT). Enter the first digit of the first data value of the instruction. Then use the (←) and (→) key as needed to cursor up to the word address. Enter the appropriate digits of the word address.

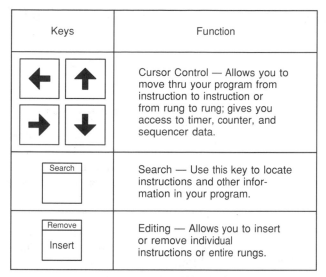

Keys	Function
← ↑ → ↓	Cursor Control — Allows you to move thru your program from instruction to instruction or from rung to rung; gives you access to timer, counter, and sequencer data.
Search	Search — Use this key to locate instructions and other information in your program.
Remove / Insert	Editing — Allows you to insert or remove individual instructions or entire rungs.

Fig. 13-7 Editing controls.

until the operator leaves the program mode.

Often a *continuous test mode* is provided which causes the processor to operate from the user program without energizing any outputs. This allows the control program to be executed and debugged while the outputs are disabled. A check of each rung can be done by monitoring the corresponding output rung on the programming device. A *single-scan* test mode may also be available for debugging the control logic. This mode causes the processor to complete a single scan of the user program each time the SINGLE SCAN key is pressed with no outputs being energized.

An *on-line programming mode* permits the user to change the program during machine operation. As the PLC controls its equipment or process, the user can add, change, or delete control instructions and data values as desired. Any modification made is executed immediately upon entry of the instruction. Therefore, the user should assess all possible sequences of machine operation resulting from the change in advance. On-line programming should be done only by experienced personnel who fully understand the operation of the PLC they are dealing with and the machinery being controlled. If at all possible, changes should be made off-line in order to provide a safe transition from existing programming to new programming.

Off-line programming. This powerful feature allows you to develop and edit processor control programs off-line (without the need of a programmable controller) by utilizing the memory of the personal computer. Programs can be developed in a comfortable (non-factory) environment, documented, stored on disk, and downloaded into the processor memory. *(Courtesy of Square D Company)*

13-8 PREVENTIVE MAINTENANCE

The biggest deterrent to PLC system faults is a proper preventive maintenance program. Although PLCs have been designed to minimize maintenance and provide trouble-free operation, there are several preventive measures that should be looked at on a regular basis.

Many control systems operate processes that must be shut down for short periods of time for product changes. The following preventive maintenance tasks should be carried out during these short shutdown periods:

- Any filters that have been installed in enclosures should be cleaned or replaced to ensure that clear air circulation is present inside the enclosure.
- Dust or dirt accumulated on PLC circuit boards should be cleaned. If dust is allowed to build up on heat sinks and electronic circuitry, an obstruction of heat dissipation could occur and cause circuit malfunction. Furthermore, if conductive dust reaches the electronic boards, a short circuit could result and cause permanent damage to the circuit board. Ensuring that the enclosure door is kept closed will prevent the rapid buildup of these contaminants.
- Connections to the I/O modules should be checked for tightness to ensure that all plugs, sockets, terminal strips, and module connections are making connections, and that the module is securely installed. Loose connections may result not only in improper function of the controller, but also in damage to the components of the system.
- All field I/O devices should be inspected to ensure that they are properly adjusted. Circuit boards dealing with process control analogs should be calibrated every 6 months. Other devices, such as sensors, should be done on a monthly basis. End devices in the environment, which have to translate mechanical signals into electrical, may gum up, get dirty, crack, or break—and then they will no longer trip at the correct setting.
- Care should be taken to ensure that heavy noise- or heat-generating equipment is not moved to close to the PLC.

13-9 TROUBLESHOOTING

In the event of a PLC fault, a careful, systematic approach should be used when troubleshooting the system to resolve the problem. The ability to monitor the activity of equipment control logic on a CRT screen is a big advantage in many troubleshooting situations.

When a problem does occur, the first step in the troubleshooting procedure is to identify the problem and its source. The source of a problem can generally be narrowed down to the processor module, I/O hardware, wiring, or machine inputs or outputs. Once a problem is recognized, it is usually quite simple to deal with. The following will deal with troubleshooting these potential problem areas:

Processor Module

The processor is responsible for the *self-detection* of potential problems. It performs error checks during its operation and sends status information to indicators that are normally located on the front of the processor module. Typical diagnostics include memory OK, processor OK, battery OK, and power supply OK.

The processor then continually monitors itself for any problems that might cause the controller to execute the user program improperly. Depending on the controller, a set of fault relay contacts may be available. The fault relay is controlled by the processor and is activated when one or more specific fault conditions occur. The fault relay contacts are used to disable the outputs and signal a failure.

Most PLCs incorporate a *watchdog timer* to monitor the scan process of the system. The watchdog timer is usually a separate timing circuit that must be set and reset by the processor within a predetermined time period. The watchdog timer circuit will *time-out* if a processor hard-wire malfunction occurs, and will immediately halt the operation of the PLC. Errors in memory data are also detected through various built-in diagnostic routines.

Input Malfunctions

If the controller is operating in the RUN mode but output devices do not operate as programmed, the most likely problem source is one of the following:

- I/O devices
- Wiring between I/O modules, I/O devices, and user power
- User power
- I/O modules

Narrowing down to one of the above as the problem source can usually be accomplished by comparing the actual status of the suspect I/O with controller status indicators. Usually each input or output device has at least two status indicators. One of these indicators is on the I/O module, the other indicator is provided by the programming device monitor.

If input hardware is suspected as being the source of a problem, the first check is to see if the status indicator on the input module illuminates when it is receiving power from its corresponding input device (e.g., push button, limit switch). If the status indicator on the input module *does not illuminate* when the input device is on, take a voltage measurement across the input terminal to check for the proper voltage level (Fig. 13-8). If the voltage

Technician monitoring a PLC system on a CRT screen.   *(Courtesy of PSI Repair Services, Inc.)*

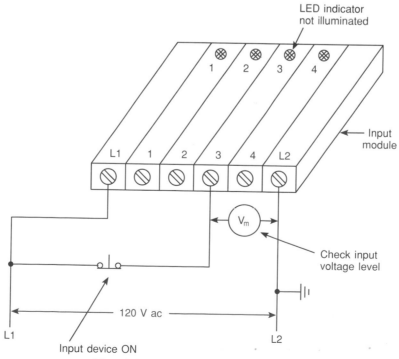

Fig. 13-8 Checking for input malfunctions.

Input device troubleshooting guide				
Input device condition	Input module status indicator	Operator terminal status indicator		Possible problem source(s)
		—‖—	—‖✗—	
Closed – ON	ON	TRUE Dark	FALSE Normal	None, correct status indication.
Open – OFF	OFF	FALSE Normal	TRUE Dark	None, correct status indication.
Closed – ON	ON	FALSE Normal	TRUE Dark	1. I/O module. 2. Processor/operator terminal communication.
Closed – ON	OFF	FALSE Normal	TRUE Dark	1. Wiring/power to I/O module. 2. I/O module.
Open – OFF	OFF	TRUE Dark	FALSE Normal	1. Programming error. 2. Processor/operator terminal communication.
Open – OFF	ON	TRUE Dark	FALSE Normal	1. Short circuit in input device or wiring. 2. Input module.

Fig. 13-9 Typical input device troubleshooting guide. *(Courtesy of Allen-Bradley Company, Inc.)*

level is correct, then the input module should be replaced. If the voltage level is not correct, then check for faults with power to the input device; wiring between input device, input module, and user power; and the input device itself.

If the programming device monitor does not show the correct status indication for a condition instruction, the input module may not be properly converting the input signal to the logic level voltage required by the processor module. In this case the input module should be replaced. If a replacement module *does not* eliminate the problem and wiring is assumed correct, then the I/O rack, communication cable, or processor should be suspected. Figure 13-9 shows a typical input device troubleshooting guide. This guide reviews condition instructions and how their TRUE/FALSE status relates to external input de-

Output device troubleshooting guide			
Output device condition	Output module status indicator	Operator terminal status indicator	Possible problem source
Energized – ON	ON	TRUE Dark	None, correct status indication.
De-energized – OFF	OFF	FALSE Normal	None, correct status indication.
De-energized – OFF	ON	TRUE Dark	1. Wiring to output device. 2. Output device.
De-energized – OFF	OFF	TRUE Dark	1. Blown fuse — output module. 2. Output module malfunction.

Fig. 13-10 Typical output device troubleshooting guide. *(Courtesy of Allen-Bradley Company, Inc.)*

vices. Status indication is provided by displaying all logically TRUE instructions in a rung in reverse video (darkened).

Output Malfunctions

When an output does not energize as expected, first check the output module blown fuse indicator. Usually this indicator will illuminate only when the output circuit corresponding to the blown fuse is energized. If this indicator is illuminated, correct the cause of the malfunction and replace the blown fuse in the module.

If the blown fuse indicator is not illuminated (fuse OK), then check to see if the output device is responding to the LED status indicator. If an output rung is energized, the module status indicator is on, and the output device is not responding, then the wiring to the output device or the output device itself should be suspected. If, according to the programming device monitor, an output device is being commanded to turn on, but the status indicator is off, then the module should be replaced. Figure 13-10 shows a typical output device troubleshooting guide.

REVIEW QUESTIONS

1. Why are PLCs generally placed within an enclosure?

2. What methods are used to keep enclosure temperatures within allowable limits?

3. State two ways in which electrical noise may be coupled into a PLC control system?

4. List four potential noise-generating devices.

5. Describe two ways in which careful wire routing can help cut down on electrical noise.

6. (a) What type of input field devices and output modules are most likely to have a small leakage current flow when they are in the OFF state? Why?
   (b) How can leakage currents be reduced?

7. When line voltage variations to the PLC power supply are excessive, what can be done to solve the problem?

8. (a) Under what condition will an inductive load generate a very high voltage spike?
   (b) What can be used to suppress a dc load?
   (c) What can be used to suppress an ac load?

9. (a) What is the purpose of PLC editing functions?
   (b) What is the purpose of the SEARCH function as part of the editing process?
   (c) What is the purpose of the cursor control keys as part of the editing process?

10. (a) Explain the difference between off-line and on-line programming.
    (b) Which method is safer? Why?

11. List four preventive maintenance tasks that should be carried out on the PLC installation on a regular basis.

12. (a) State two important reasons for properly grounding a PLC installation.
    (b) List four important grounding practices to follow when installing a PLC system.

13. (a) List four types of diagnostic fault indicators that often operate from the processor's self-detection circuits.
    (b) When a processor comes equipped with a fault relay, how does this circuit usually operate?
    (c) What is the prime function of a watchdog timer circuit?

PROBLEMS

1. The enclosure door of a PLC installation is not kept closed. What potential problem could this create?

2. A fuse is blown in an output module. Suggest three possible reasons why the fuse blew.

3. Whenever a crane located over a PLC installation is started from a standstill, temporary malfunction of the PLC system is experienced. What is one likely cause of the problem?

4. During the static checkout of a PLC system, a specific output is forced on by the programming device. If an indicator other than the expected one turns on, what is the probable problem?

5. The input device to a module is activated, but the LED status indicator does not come on. A check of the voltage to the input module indicates no voltage is present. Suggest two possible causes of the problem.

6. An output is forced on. The module logic light comes on, but the field device does not work. A check of the voltage on the output module indicates the proper voltage level. Suggest two possible causes of the problem.

7. A specific output is forced on, but the LED module indicator does not come on. A check of the voltage at the output module indicates a voltage far below the normal ON level. What is the first thing to check?

8. An electronic-based input sensor is wired to a high-impedance PLC input and is falsely activating the input. How can this problem be corrected?

9. An LED logic indicator is illuminated, and according to the programming device monitor, the processor is not recognizing the input. If a replacement module does not eliminate the problem, what two other things should be suspected?

10. (a) An NO field limit switch examined for an ON state normally cycles from ON to OFF five times during one machine cycle. How could you tell by observing the LED status light that the limit switch is functioning properly?

(b) How could you tell by observing the programming device monitor that the limit switch is functioning properly?

(c) How could you tell by observing the LED status light whether the limit switch was stuck open?

(d) How could you tell by observing the programming device monitor whether the limit switch was stuck open?

(e) How could you tell by observing the LED status light if the limit switch was stuck closed?

(f) How could you tell by observing the programming device monitor if the limit switch was stuck closed?

11. Assume that prior to putting a PLC system into operation, you want to verify that each *input device* is connected to the correct input terminal and that the input module or point is functioning properly. Outline the safest method of carrying out this test.

12. Assume that prior to putting a PLC system into operation, you want to verify that each *output device* is connected to the correct output terminal and that the output module or point is functioning properly. Outline the safest method of carrying out this test.

Appendix: Graphic Symbols and Charts

Typical graphic symbols for electrical diagrams

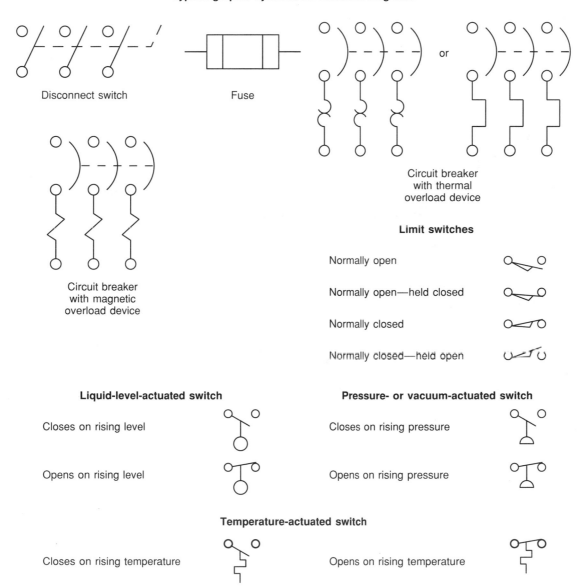

Disconnect switch

Fuse

or

Circuit breaker
with thermal
overload device

Circuit breaker
with magnetic
overload device

Limit switches

Normally open

Normally open—held closed

Normally closed

Normally closed—held open

Liquid-level-actuated switch

Closes on rising level

Opens on rising level

Pressure- or vacuum-actuated switch

Closes on rising pressure

Opens on rising pressure

Temperature-actuated switch

Closes on rising temperature

Opens on rising temperature

Flow-actuated switch

Closes on increase in flow   Opens on increase in flow

Foot-operated switch   **Selector switch**

Opens by foot pressure

Closes by foot pressure

Rotary, break-before-make, nonshorting   Rotary, make-before-break, shorting
(non-bridging) during contact transfer   (bridging) during contact transfer

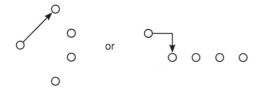   or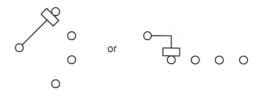

Rotary, segmental contact

   or

Motors

MTR

3 Phase

A

DC motor

Contacts

Normally open

Normally closed

Normally open with time delay closing (TC)

Normally open with time delay opening (TO)

Normally closed with time delay opening (TO)

Normally closed with time delay closing (TC)

Time sequential closing

Static relay (proximity switch)

Normally ON

Normally OFF

**Pushbutton, momentary or
spring-return**

Circuit closing (make)

Circuit opening (break)

Two-circuit

Mushroom-head
(shown with 2-circuit pushbutton
switch)

Push button, two-circuit, maintained or not
spring-return

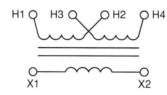

Coils

Relay

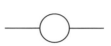

Solenoid

Thermal element actuating device

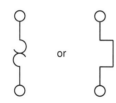

or

Thermocouple
(temperature measuring device)

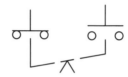

Control circuit transformer

H1   H3   H2   H4

X1   X2

Numbering system conversion chart

Hexadecimal	Binary	Decimal	Octal
0	0000	0	000
1	0001	1	001
2	0010	2	002
3	0011	3	003
4	0100	4	004
5	0101	5	005
6	0110	6	006
7	0111	7	007
8	1000	8	010
9	1001	9	011
A	1010	10	012
B	1011	11	013
C	1100	12	014
D	1101	13	015
E	1110	14	016
F	1111	15	017

Gate symbols and truth tables

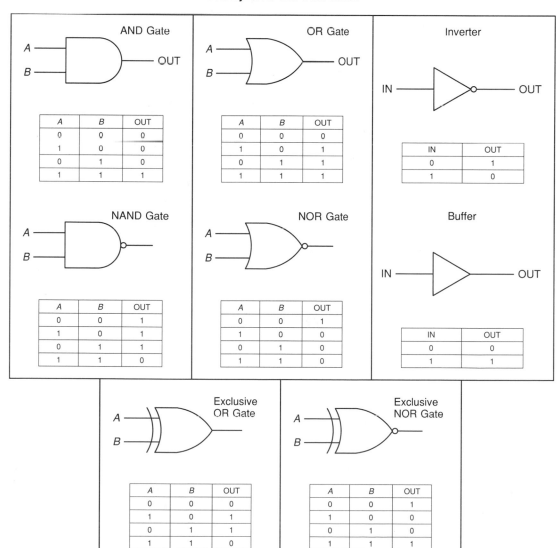

AND Gate

A	B	OUT
0	0	0
1	0	0
0	1	0
1	1	1

OR Gate

A	B	OUT
0	0	0
1	0	1
0	1	1
1	1	1

Inverter

IN	OUT
0	1
1	0

NAND Gate

A	B	OUT
0	0	1
1	0	1
0	1	1
1	1	0

NOR Gate

A	B	OUT
0	0	1
1	0	0
0	1	0
1	1	0

Buffer

IN	OUT
0	0
1	1

Exclusive OR Gate

A	B	OUT
0	0	0
1	0	1
0	1	1
1	1	0

Exclusive NOR Gate

A	B	OUT
0	0	1
1	0	0
0	1	0
1	1	1

GLOSSARY

A

Access To locate data stored in a programmable logic controller system or in computer related equipment.

Accumulated value The number of elapsed timed intervals or counted events.

Address A code that indicates the location of data to be used by a program, or the location of additional program instructions.

Alphanumeric Term describing character strings consisting of any combination of alphabets, numerals, and/or special characters (e.g., A15$) used for representing text, commands, numbers, and/or code groups.

Alternating current (ac) input module An input/output module that converts various alternating current signals originating at user devices to the appropriate logic level signal for use within the processor.

Alternating current (ac) output module An input/output module that converts the logic level signal of the processor to a usable output signal to control a user alternating current device.

Ambient temperature The temperature of the air surrounding a module or system.

American National Standard Code for Information Interchange (ASCII) An 8-bit (7 bits plus parity) code that represents all characters of a standard typewriter keyboard, both uppercase and lowercase, as well as a group of special characters that are used for control purposes.

American National Standards Institute (ANSI) A clearinghouse and coordinating agency for voluntary standards in the United States.

American wire gauge (AWG) A standard system used for designating the size of electrical conductors. Gauge numbers have an inverse relationship to size; larger numbers have a smaller diameter.

Analog device Apparatus that measures continuous information (e.g., voltage-current). The measured analog signal has an infinite number of possible values. The only limitation on resolution is the accuracy of the measuring device.

Analog input module An input circuit that employs an analog-to-digital converter to convert an analog value, measured by an analog measuring device, to a digital value that can be used by the processor.

Analog output module An output circuit that employs a digital-to-analog converter to convert a digital value, sent from the processor, to an analog value that will control a connected analog device.

Analog signal Signal having the characteristic of being continuous and changing smoothly over a given range, rather than switching suddenly between certain levels as with discrete signals.

AND (logic) A boolean operation that yields a logic 1 output if all inputs are 1, and a logic 0 if any input is 0.

Arithmetic capability The ability to do addition, subtraction, multiplication, division, and other advanced math functions with the processor.

Auxiliary power supply A power supply that is not associated with the processor. Auxiliary power supplies are usually required to supply logic power to input/output racks and to other processor support hardware, and are often referred to as *remote power supplies*.

B

Backplane A printed circuit board, located in the back of a chassis, that contains a data bus, power bus, and mating connectors for modules to be inserted in the chassis.

BASIC A computer language using brief English-language statements to instruct a computer or microprocessor.

Battery indicator A diagnostic aid that provides a visual indication to the user, and/or an internal processor software indication, that the memory power-fail support battery is in need of replacement.

Binary A number system using 2 as a base. The binary number system requires only two digits, zero (0) and one (1), to express any alphanumeric quantity desired by the user.

Binary-coded decimal (BCD) A system of numbering that expresses each individual decimal digit (0 through 9) of a number as a series of 4-bit binary notations. The binary-coded decimal system is often referred to as 8421 code.

Binary word A related group of 1's and 0's that has meaning assigned by position, or by numerical value in the binary system of numbers.

Bit An acronym for the words *binary digit*. The bit is the smallest unit of information in the binary numbering system. It represents a decision between one of two possible and equally likely values or states. It is often used to represent an OFF or ON state as well as a TRUE or FALSE condition.

Bit manipulation instructions A family of programmable logic controller instructions which exchange, alter, move, or otherwise modify the individual bits of single or groups of processor data memory words.

Bit storage A user-defined data table area in which bits can be set or reset without directly affecting or controlling output devices. However, any storage bit can be monitored as necessary in the user program.

Block diagram A method of representing the major functional subdivisions, conditions, or operations of an overall system, function, or operation.

Block transfer An instruction that copies the contents of one or more contiguous data memory words to a second contiguous data memory location. An instruction that transfers data between an intelligent input/output module or card and specified processor data memory locations.

Boolean algebra A mathematical shorthand notation that expresses logic functions, such as AND, OR, EXCLUSIVE OR, NAND, NOR, and NOT.

Branch A parallel logic path within a rung.

Buffer In software terms, a register or group of registers used for temporary storage of data; a buffer is used to compensate for transmission rate differences between the transmitter and receiving device. In hardware terms, an isolating circuit used to avoid the reaction of one circuit with another.

Burn The process by which information is entered into programmable read-only memory.

Bus A group of lines used for data transmission or control. Power distribution conductors.

Byte A group of adjacent bits usually operated upon as a unit, such as when moving data to and from memory. There are 8 bits per byte.

C

Cascading A technique used when programming timers and counters to extend the timing or counting range beyond what would normally be available. This technique involves the driving of one timer or counter instruction from the output of another similar instruction.

Central processing unit (CPU) That part of the programmable logic controller that governs system activities, including the interpretation and execution of programmed instructions. The central processing unit is also referred to as the *processor* or the *CPU*.

Character A symbol that is one of a larger group of similar symbols and that is used to represent information on a display device. The letters of the alphabet and the decimal numbers are examples of characters used to convey information.

Chassis A housing or framework that is used to hold assemblies. When the chassis is filled with one or more assemblies, it is often referred to as a *rack*.

Chip A very small piece of semiconductor material, on which electronic components are formed. Chips are normally made of silicon and are typically less than 1/4 in square and 1/100 in thick.

CLEAR An instruction or a sequence of instructions which removes all current information from a programmable logic controller's memory.

Coaxial cable A transmission line that is constructed such that an outer conductor forms a cylinder around a central conductor. An insulating dielectric separates the inner and outer conductors, and the complete assembly is enclosed in a protective outer sheath. Coaxial cables are not susceptible to external electric and magnetic fields and generate no electric or magnetic fields of their own.

Code A system of communications that uses arbitrary groups of symbols to represent information or instructions. Codes are usually employed for brevity or security.

COMPARE An instruction that compares for equality or inequality the contents of two designated data memory locations of a programmable logic controller.

Compatibility The ability of various specified units to replace one another with little or no reduction in capability. The ability of units to be interconnected and used without modification.

Complementary metal-oxide semiconductor (CMOS) CMOS-based logic offers lower power consumption and high-speed operation.

Computer Any electronic device which is able to accept information, manipulate it according to a set of preprogrammed instructions, and supply the results of the manipulation.

Computer interface A device designed for data communication between a programmable logic controller and a computer.

Contact The current-carrying part of an electric relay or switch; the contact engages to permit power flow and disengages to interrupt power flow to a load device.

Contact bounce The uncontrollable making and breaking of a contact during the initial engaging or disengaging of the contact.

Contact histogram An instruction sequence that monitors a designated memory bit or a designated input or output point for a change of state. A listing is generated by the instruction sequence that displays how quickly the monitored point is changing state.

Contactor A special-purpose relay that is designed to establish and interrupt the power flow of high-current electric circuits.

Contact symbology A set of symbols used to express the control program using conventional relay symbols.

Control logic The control plan for a given system. The program.

Control relay A relay used to control the operation of an event or a sequence of events.

Core memory A type of memory system that employs ferrite cores to store information. Core memory operates by magnetizing the ferrite core in one direction to represent a 1, ON, or TRUE state, and in the opposite direction to represent a 0, OFF, or FALSE state. This form of memory is nonvolatile.

Counter An electromechanical device in relay-based control systems that counts numbers of events for the purpose of controlling other devices based on the current number of counts recorded. A programmable logic controller instruction that performs the functions of its electromechanical counterpart.

Cathode-ray tube (CRT) terminal A portable enclosure containing a cathode-ray tube, a special-purpose keyboard, and a microprocessor which is used to program a programmable logic controller.

Current The rate of electrical electron movement, measured in amperes.

Current-carrying capacity The maximum amount of current a conductor can carry without heating beyond a predetermined safe limit.

Current sink A device that receives current.

Current source A device that supplies current.

Cursor The intensified or blinking element in the user program or file display. A means for indicating on the screen of a cathode-ray tube the point where data entry or editing occurs.

Cycle A sequence of operations that is repeated regularly. The time it takes for one such sequence to occur.

D

Data Information encoded in a digital form, which is stored in an assigned address of data memory for later use by the processor.

Data address A location in memory where data can be stored.

Data file A group of data memory words which are acted upon as a group rather than singly.

Data link The equipment that makes up a data communications network.

Data manipulation The process of exchanging, altering, or moving data within a programmable logic controller or between programmable logic controllers.

Data manipulation instructions A classification of processor instructions that alter, exchange, move, or otherwise modify data memory words.

Data table The part of processor memory that contains input and output values as well as files where data is monitored, manipulated, and changed for control purposes.

Data transfer The process of moving information from one location to another, i.e., from register to register, from device to device, and so forth.

Data transmission line A medium for transferring signals over a distance.

Debouncing The act of removing intermediate noise states from a mechanical switch.

Debug The process of locating and removing mistakes from a software program or from hardware interconnections.

Decimal number system A number system that uses ten numeral digits (decimal digits), 0, 1, 2, 3, 4, 5, 6, 7, 8, 9. Each digit position has a place value of 1, 10, 100, 1000, and so on, beginning with the least significant (rightmost) digit. Base 10.

Decrement The act of reducing the contents of a storage location or value in varying increments.

Diagnostics Pertains to the detection and isolation of an error or malfunction.

Diagnostic program A user program designed to help isolate hardware malfunctions in the programmable logic controller and the application equipment.

Digital device One that processes discrete electric signals.

Digital signal Signal having the characteristic of being discrete or discontinuous in nature. One that is present or not present, can be counted and represented directly as a numerical value.

Discrete I/O A group of input and/or output modules that operate with ON/OFF signals as contrasted to analog modules that operate with continuously variable signals.

Disk drive The device that writes or reads data from a magnetic disk.

Display The image which appears on a cathode-ray tube screen or on other image projection systems.

Display menu The list of displays from which the user selects specific information for viewing.

DIVIDE A programmable logic controller instruction which performs a numerical division of one number by another.

Documentation An orderly collection of recorded hardware and software data such as tables, listings, and diagrams to provide reference information for programmable logic controller application operation and maintenance.

E

Edit The act of modifying a programmable logic controller program to eliminate mistakes and/or simplify or change system operation.

Electrically erasable programmable read-only memory (EEPROM) A type of programmable read-only memory that is programmed and erased by electrical pulses.

Electrical-optical isolator A device which provides electrical isolation using a light source and detector in the same package.

Electromagnetic interference (EMI) A phenomenon responsible for noise in electric circuits.

Element A single instruction of a relay ladder diagram program.

Emergency STOP relay A relay used to inhibit all electric power to a control system in the event of an emergency or other event requiring that the controlled hardware be brought to an immediate halt.

Enable To permit a particular function or operation to occur under natural or preprogrammed conditions.

Enclosure A steel box with a removable cover or hinged door, used to house electric equipment.

Encoder A rotary device which transmits position information. A device which transmits a fixed number of pulses for each revolution.

Energize The physical application of power to a circuit or device in order to activate it. The act of setting the ON, TRUE, or 1 state of a programmable logic controller's relay ladder diagram output device or instruction.

Erasable programmable read-only memory (EPROM) A programmable read-only memory that can be erased with ultraviolet light, then reprogrammed with electrical pulses.

Error signal A signal proportional to the difference between the actual output and the desired output.

EXAMINE OFF Refers to a normally closed contact instruction in a logic ladder program. An EXAMINE OFF instruction is TRUE if its addressed bit is OFF (0). It is FALSE if the bit is ON (1).

EXAMINE ON Refers to a normally open contact instruction in a logic ladder program. An EXAMINE ON instruction is TRUE if its addressed bit is on (1). It is FALSE if the bit is OFF (0).

Execution The performance of a specific operation that is accomplished through processing one instruction, a series of instructions, or a complete program.

Execution time The total time required for the execution of one specific operation.

F

FALSE As related to programmable logic controller instructions, a disabling logic state.

Fault Any malfunction which interferes with normal operation.

Fault indicator A diagnostic aid that provides a visual indication and/or an internal processor software indication that a fault is present in the system.

File A formatted block of data that is treated as a unit.

Flow chart A graphical representation for the definition, analysis, or solution of a problem. Symbols are used to represent a process or sequence of decisions and events.

Force OFF function A feature which allows the user to reset an input image table bit or de-energize an output, independent of the programmable logic controller program.

FORCE ON function A feature which allows the user to set an image table bit or energize an output, independent of the programmable logic controller program.

Functional block instruction set A set of instructions that moves, transfers, compares, or sequences blocks of data.

G

Gate A circuit having two or more input terminals and one output terminal, where an output is present when and only when the prescribed inputs are present.

GET instruction A programmable logic controller instruction that fetches the contents of a specified data memory word. GET instructions are often used to fetch data prior to programming mathematical and data manipulation instructions.

Gray code A binary coding scheme that allows only 1 bit in the data word to change state at each increment of the code sequence.

Ground A conducting connection between an electric circuit or equipment chassis and the earth ground.

Ground potential Zero voltage potential with respect to the ground.

H

Hard contacts Any type of physical switch contacts.

Hard copy Any form of a printed document such as a ladder diagram program listing, paper tape, or punched cards.

Hardware The mechanical, electric, and electronic devices that make up a programmable logic controller and its application.

Hard-wired Describes the physical interconnection of electric and electronic components with wire.

Hexadecimal A number system having a base of 16. This numbering system requires 16 elements for representation, and thus uses the decimal digits zero (0) through nine (9) and the first six letters of the alphabet, A through F.

High-level language A powerful set of user-oriented instructions in which each statement may translate into a series of instructions or subroutines in machine language.

HIGH = TRUE A signal type where the higher of two voltages indicates a logic state of ON (1).

Histogram A graphic representation of the frequency at which an event occurs.

I

Image table An area in programmable logic controller memory dedicated to input/output data. Ones and zeros (1 and 0) represent ON and OFF conditions, respectively. During every input/output scan, each input controls a bit in the input image table; each output is controlled by a bit in the output image table.

IMMEDIATE INPUT instruction A programmable logic controller instruction that temporarily halts the user program scan so that the processor can update the input image table with the current status of one or more user-specified input points.

IMMEDIATE OUTPUT instruction A programmable logic controller instruction that temporarily halts the user program scan so that the current status of one or more user-specified output points can be updated to current output image table statuses by the processor.

Impedance The total resistive and inductive opposition that an electric circuit or device offers to a varying current at a specified frequency. Impedance is measured in ohms and is denoted by the symbol Z.

Increment The act of increasing the contents of a storage location or value in varying amounts.

Inductance A circuit property that opposes any current change. Inductance is measured in henrys and is represented by the letter H.

Input Information transmitted from a peripheral device to the input module, and then to the data table.

Input devices Devices such as limit switches, pressure switches, push buttons, analog and/or digital devices, that supply data to a programmable logic controller.

Instruction A command that causes a programmable logic controller to perform one specific operation. The user enters a combination of instructions into the programmable logic controller's memory to form a unique application program.

Instruction set The set of general-purpose instructions available with a given controller. In general, different machines have different instruction sets.

Intelligent input/output module A microprocessor-based module that performs processing or sophisticated closed-loop application functions.

Interface A circuit that permits communication between the central processing unit and a field input or output device. Different devices require different interfaces.

Internal coil instruction A relay coil instruction used for internal storage or buffering of an ON/OFF logic state. An internal coil instruction differs from an output coil instruction in that the ON/OFF status of the internal coil is not passed to the input/output hardware for control of a field device.

Input/output (I/O) address A unique number assigned to each input and output. The address number is used when programming, monitoring, or modifying a specific input or output.

Input/output (I/O) module A plug-in type assembly that contains more than one input or output circuit. A module usually contains two or more identical circuits. Normally it contains 2, 4, 8, or 16 circuits.

Input/output (I/O) update The continuous process of revising each and every bit in the input and output tables, based on the latest results from reading the inputs and processing the outputs according to the control program.

Input/output (I/O) scan time The time required for the processor to monitor inputs and control outputs.

Isolated input/output (I/O) circuits Input and output circuits that are electrically isolated from any and all other circuits of a module. Isolated input/output circuits are designed to allow field devices that are powered from different sources to be connected to one module.

J

Jumper A short length of conduit used to make a connection between terminals around a break in a circuit.

JUMP instruction An instruction that permits the bypassing of selected portions of the user program. JUMP instructions are conditional whenever their operation is determined by a set of preconditions, and unconditional whenever they are executed to occur every time programmed.

K

K $2^{10} = 1K = 1024$. Used to denote size of memory and can be expressed in bits, bytes, or words. Example: 2K = 2048.

k Kilo. A prefix used with units of measurement to designate quantities 1000 times as great.

Keying Keying bands installed on backplane connectors to ensure that only one type of module can be inserted into a keyed connector.

L

LABEL instruction A programmable logic controller instruction that assigns an alphanumeric designation to a particular location in a program. This location is used as the target of a JUMP, SKIP, or JUMP TO SUBROUTINE instruction.

Ladder diagram An industry standard for representing relay logic control systems. The diagram resembles a ladder in that the vertical supports of the ladder appear as power feed and return buses, and the horizontal rungs of the ladder appear as series and/or parallel circuits connected across the power lines.

Ladder diagram programming A method of writing a user program in a format similar to a relay ladder diagram.

Ladder matrix A rectangular array of programmed contacts that defines the number of contacts that can be programmed across a row and the number of parallel branches allowed in a single ladder rung.

Language A set of symbols and rules for representing and communicating information among people, or between people and machines. The method used to instruct a programmable device to perform various operations. Examples include boolean and ladder contact symbology.

Latching relay A relay that maintains a given position by mechanical or electrical means until released mechanically or electrically.

LATCH instruction One-half of an instruction pair (the second instruction of the pair being the UNLATCH instruction) which emulates the latching action of a latching relay. The LATCH instruction for a programmable logic controller energizes a specified output point or internal coil until it is de-energized by a corresponding UNLATCH instruction.

Leakage The small amount of current that flows in a semiconductor device when it is in the OFF state.

Least significant bit (LSB) The bit that represents the smallest value in a nibble, byte, or word.

Least significant digit (LSD) The digit that represents the smallest value in a byte or word.

Light-emitting diode (LED) A semiconductor PN-type junction that emits light when biased in the forward direction.

Light-emitting diode (LED) display A display device incorporating light-emitting diodes to form the segments of the displayed characters and numbers.

Limit switch An electric switch actuated by some part and/or motion of a machine or equipment.

Line A component part of a system used to link various subsystems located remotely from the processor. The source of power for operation. Example: 120V alternating current line.

Liquid-crystal display (LCD) A display device using reflected light from liquid crystals to form the segments of the displayed characters and numbers.

Load The power used by a machine or apparatus. To place data into an internal register under program control. To place a program from an external storage device into central memory under operator control.

Load resistor A resistor connected in parallel with a high-impedance load so that the output circuit driving the load can provide at least the minimum current required for proper operation.

Local input/output (I/O) A programmable logic controller whose input/output distance is physically limited. The PLC must be located near the processor; however, the PLC may still be mounted in a separate enclosure.

Local power supply The power supply used to provide power to the processor and a limited number of local input/output modules.

Location In reference to memory, a storage position or register identified by a unique address.

Logic A process of solving complex problems through the repeated use of simple functions that can be either TRUE or FALSE. The three basic logic functions are AND, OR, and NOT.

Logic diagram A diagram which represents the logic elements and their interconnections.

Logic level The voltage magnitude associated with signal pulses representing 1's and 0's in binary computation.

Loop resistance The total resistance of two conductors measured at one end (conductor and shield, twisted pair, conductor and armor).

LOW = TRUE A signal type where the lower of two voltages indicates a logic state of ON (1).

M

Machine language A programming language using the binary form.

Magnetic disk A flat, circular plate with a magnetic surface on which data can be stored by selective polarization.

Magnetic tape Tape made of plastic and coated with magnetic material; used to store information.

Malfunction Any incorrect function within electronic, electric, or mechanical hardware.

Manipulation The process of controlling and monitoring data table bits, bytes, or words by means of the user program to vary application functions.

Masking A means of selectively screening out data. Masking allows unused bits in a specific instruction to be used independently.

Master control relay (MCR) A mandatory hard-wired relay that can be de-energized by any series-connected emergency STOP switch. Whenever the master control relay is de-energized, its contacts open to de-energize all application input and output devices.

Master control relay (MCR) zones User program areas where all nonretentive outputs can be turned off simultaneously. Each master control relay zone must be delimited and controlled by master control relay fence codes (master control relay instructions).

Memory That part of the programmable logic controller where data and instructions are stored either temporarily or semipermanently. The control program is stored in memory.

Memory map A diagram showing a system's memory addresses and what programs and data are assigned to each section of memory.

Metal-oxide semiconductor (MOS) A semiconductor device in which an electric field controls the conductance of a channel under a metal electrode called a *gate*.

Microprocessor A central processing unit that is manufactured on a single integrated-circuit chip (or several chips) by utilizing large-scale integration technology.

Microsecond One millionth of a second = 1×10^{-6} second = 0.000001 second.

Millisecond One thousandth of a second = 1×10^{-3} second = 0.001 second.

Mnemonic A term, usually an abbreviation, that is easy to remember.

Mnemonic code A code in which information is represented by symbols or characters.

Mode A term used to refer to the selected operating method such as automatic, manual, TEST, PROGRAM, or diagnostic.

Module An interchangeable, plug-in item containing electronic components.

Module addressing A method of identifying the input/output modules installed in a chassis.

Module group Two or more modules which as a group perform a specific function or operation, or are thought of as a single unit.

Monitor Any display device incorporating a cathode-ray tube as the primary display medium. The act of listening to or observing the operation of a system or device.

Most significant bit (MSB) The bit representing the greatest value of a nibble, byte, or word.

Most significant digit (MSD) The digit representing the greatest value of a byte or word.

Motor controller or starter A device or group of devices that serve to govern, in a predetermined manner, the electric power delivered to a motor.

Multiplexing The time-shared scanning of a number of data lines into a single channel. Only one data line is enabled at any time. The incorporation of two or more signals into a single wave from which the individual signals can be recovered.

MULTIPLY instruction A programmable logic controller instruction that provides for the mathematical multiplication of two numbers.

N

National Electrical Code (NEC) A set of regulations developed by the National Fire Protection Association which govern the construction and installation of electric wiring and electric devices. The National Electrical Code is recognized by many governmental bodies, and compliance is mandatory in much of the United States.

National Electrical Manufacturers Association (NEMA) An organization of electric device and product manufacturers. The National Electrical Manufacturers Association issues standards relating to the design and construction of electric devices and products.

NEMA Type 12 enclosure A category of industrial enclosures intended for indoor use and designed to provide a degree of protection against dust, falling dirt, and dripping noncorrosive liquids. They do not provide protection against conditions such as internal condensation.

Network A series of stations or devices connected by some type of communications medium.

Node In hardware, a connection point on the network. In programming, the smallest possible increment in a ladder diagram.

Noise Random, unwanted electric signals, normally caused by radio waves or electric or magnetic fields generated by one conductor and picked up by another.

Noise filter or suppressor An electronic filter network used to reduce and/or eliminate any noise that may be present on the leads to an electric or electronic device.

Noise immunity A measure of insensitivity of an electronic system to noise.

Noise spike A short burst of electric noise with more magnitude than the background noise level.

Nonretentive output An output which is continuously controlled by a program rung. Whenever the rung changes state (TRUE or FALSE), the output turns on or off. Contrasted with a retentive output, which remains in its last state (ON or OFF) depending on which of its two rungs, LATCH or UNLATCH, was last TRUE.

Nonvolatile memory A memory that is designed to retain its data while its power supply is turned off.

NOT A logical operation that yields a logic 1 at the output if a logic 0 is entered at the input, and a logic 0 at the output if a logic 1 is entered at the input. The NOT, also called the *inverter*, is normally used in conjunction with the AND and OR functions.

O

Octal numbering system A numbering system that uses only the digits 0 through 7. Also called base 8.

Off-delay timer An electromechanical relay that has contacts which change state a predetermined time period after power is removed from its coil; upon re-energization of the coil, the contacts return to their shelf state immediately. A programmable logic controller instruction that emulates the operation of the electromechanical off-delay relay.

Off-line Equipment or devices that are not connected to, or do not directly communicate with, the central processing unit.

Off-line programming and/or editing A method of programmable logic controller programming and/or editing where the operation of the processor is stopped and all output devices are switched off. Off-line programming is the safest manner to develop or edit a programmable logic controller program since the entry of instructions does not affect operating hardware until the program can be verified for accuracy of entry.

On-delay timer An electromechanical relay that has contacts which change state a predetermined time period after the coil is energized; the contacts return to their shelf state immediately upon de-energization of the coil. A programmable logic controller instruction that emulates the operation of the electromechanical on-delay timer.

One-shot A programmed technique which sets a storage bit or output for only one program scan.

On-line Equipment or devices which communicate with the device they are connected to.

On-line data change Allows the user to change various data table values using a peripheral device while the application is operating normally.

On-line programming and/or editing The ability of a processor and programming terminal to jointly make user-directed additions, deletions, or changes to a user program while the processor is actively solving and executing the commands of the existing user program. Extreme care should be exercised when performing on-line programming to ensure that erroneous system operation does not result.

Optical coupler A device that couples signals from one circuit to another by means of electromagnetic radiation, usually infrared or visible. A typical optical coupler uses a light-emitting diode to convert the electric signal of the primary circuit into light, and a phototransistor in the secondary to reconvert the light back into an electric signal. Sometimes referred to as optical isolation.

Optical isolation Electrical separation of two circuits with the use of an optical coupler.

OR A logical operation that yields a logic 1 output if one of any number of inputs is 1, and a logic 0 if all inputs are 0.

Output Information sent from the processor to a connected device via some interface. The information could be in the form of control data that will signal some device such as a motor to switch on or off, or vary the speed of a drive.

Output device Any connected equipment that will receive information or instructions from the central processing unit, such as control devices (e.g., motors, solenoids, alarms) or peripheral devices (e.g., line printers, disk drives, displays). Each type of output device has a unique interface to the processor.

Output image table A portion of a processor's data memory reserved for the storage of output device statuses. A 1, ON, or TRUE state in an output image table storage location is used to switch on the corresponding output point.

OUTPUT instruction The term applied to any programmable logic controller instruction that is capable of controlling the discrete or analog status of an output device connected to the programmable logic controller.

Output register, or output word A particular word in a processor's output image table where numerical data are placed for transmission to a field output device.

Overflow A condition that occurs whenever a data storage location used for a mathematical operation is insufficient to hold the result.

Overload A load greater than a component or system is designed to handle.

Overload relay A special-purpose relay designed such that its contacts transfer whenever its current exceeds a predetermined value. Overload relays are used with electric motors to prevent motor burnout due to mechanical overload.

P

Parallel circuit A circuit in which two or more of the connected components or contact symbols in a ladder program are connected to the same pair of terminals so that current may flow through all the branches. Contrasted with a series connection, where the parts are connected end to end so that current flow has only one path.

Parallel instruction A programmable logic controller instruction used to begin and/or end a parallel branch of instructions being programmed on a programming terminal.

Parallel operation A type of information transfer where all bits, bytes, or words are handled simultaneously.

Parity The use of a self-checking code employing binary digits in which the total number of 1's is always even or odd.

Peripheral equipment Units which communicate with the programmable logic controller, but are not part of the programmable logic controller. Example: a programming device or computer.

Pilot-type device A device used in a circuit as a control apparatus to carry electric signals for directing performance. This device does not carry primary current.

Polarity The directional indication of electrical flow in a circuit. The indication of charge as either positive or negative, or the indication of a magnetic pole as either north or south.

Port A connector or terminal strip used to access a system or circuit. Generally ports are used for the connection of peripheral equipment.

Positive logic The use of binary logic in such a way that 1 represents a positive logic level (e.g., $1 = +5V$, $0 = 0V$). This is the conventional use of binary logic.

Power supply A device used to convert an alternating current or direct current voltage of specific value to one or more direct current voltages of a specified value and current capacity. The power supplies designed for use with programmable logic controllers convert 120 or 240V alternating current to the direct current voltages necessary to operate the processor and input/output hardware.

Preset value (PR) The number of time intervals or events to be counted.

Pressure switch A switch that is activated at a specified pressure.

Printed circuit board A glass-epoxy card with copper foils for electric conductors and electronic components.

Process A continuous manufacturing operation.

Proportional-integral derivative (PID) A mathematical formula that provides a closed-loop control of a process. Inputs and outputs are continuously variable and typically will be analog signals.

Proximity switch An input device that senses the presence or absence of a target without physical contact.

Pulse A short change in the value of a voltage or current level. A pulse has a definite rise and fall time and a finite duration.

PUT instruction A programmable logic controller instruction that places the data retrieved by a GET instruction in a specified data memory location specified as part of the PUT instruction.

R

Rack A housing or framework that is used to hold assemblies. A plastic and/or metal assembly that supports input/output modules and provides a means of supplying power and signals to each input/output module or card.

Rack fault A red diagnostic indicator that lights to signal a loss of communication between the processor and any remote input/output chassis. The condition which is based on the loss of communication.

Random-access memory (RAM) A memory system that permits the random accessing of any storage location for the purpose of either storing (writing) or retrieving (reading) information. Random-access memory systems allow the data to be retrieved and stored at speeds independent of the storage locations being accessed.

Read The accessing of information from a memory system or data storage device. The gathering of information from an input device or devices or a peripheral device.

Read-only memory (ROM) A permanent memory structure where data are placed in the memory's storage locations at time of fabrication, or by the user at a speed much slower than it will be read. Information entered in a read-only memory is usually never changed once entered.

Read/write memory A memory where data can be stored (WRITE mode) or accessed (READ mode). The WRITE mode replaces previously stored data with current data; the READ mode does not alter stored data.

Real-time clock (RTC) A device that continually measures time in a system without respect to what tasks the system is performing.

Rectifier A solid-state device that converts alternating current to pulsed direct current.

Register A memory word or area used for temporary storage of data used within mathematical, logical, or transferral functions.

Relay An electrically operated device that mechanically switches electric circuits.

Relay contacts The contacts of a relay which are either opened and/or closed according to the condition of the relay coil. Relay contacts are designated as either normally open or normally closed in design.

Relay logic A representation of the program or other logic in a form normally used for relays.

Remote input/output (IO) system Any input/output system that permits communications between the processor and input/output hardware over a coaxial or twin axial cable. Remote input/output systems permit the placement of input/output hardware any distance from the processor.

Report An application data display or printout containing information in a user-designed format. Reports could include operator messages, part records, and production lists. Initially entered as messages, reports are stored in a memory area separate from the user program.

Report generation The printing or displaying of user-formatted application data by means of a data terminal. Report generation can be initiated by means of either a user program or a data terminal keyboard.

Response time The amount of time required for a device to react to a change in its input signal, or to a request.

Retentive instruction Any programmable logic controller instruction that does not need to be continuously controlled for operation. Loss of power to the instruction does not halt execution or operation of the instruction.

Retentive timer An electromechanical relay that accumulates time whenever the device receives power, and maintains the current time should power be removed from the device. Loss of power to the device after reaching its preset value does not affect the state of the contacts.

Retentive timer instruction A programmable logic controller instruction that emulates the timing operation of the electromechanical retentive timer.

Retentive timer RESET instruction A programmable logic controller instruction that emulates the reset operation of the electromechanical retentive timer.

Reverse video A cathode-ray tube display characterized by black alphanumeric characters on a white background as contrasted to the standard display of white alphanumeric characters on a black background.

Routine A series of instructions that performs a specific function or task.

Run The single continuous execution of a program by a programmable logic controller.

Rung A group of programmable logic controller instructions which controls an output or storage bit, or performs other control functions such as file moves, arithmetic, and/or sequencer instructions. This is represented as one section of a logic ladder diagram.

S

Scan time The time required to read all inputs, execute the control program, and update local and remote input and output statuses. This is effectively the time required to activate an output that is controlled by programmed logic.

Schematic A diagram of graphic symbols representing the electrical scheme of a circuit.

Screen The viewing surface of a cathode-ray tube, where data is displayed.

SEARCH function Allows the user to quickly display any instruction in the programmable logic controller program.

Self-diagnostic The hardware and firmware within a controller that monitors its own operation and indicates any fault which it can detect.

Sensor A device used to gather information by the conversion of a physical occurrence to an electric signal.

Sequencer A mechanical, electric, or electronic device that can be programmed so that a predetermined set of events occurs repeatedly.

Serial operation A type of information transfer where the bits are handled sequentially. Contrasted with parallel operation.

Series circuit A circuit in which the components or contact symbols are connected end to end, and all must be closed to permit current flow.

Shield A barrier, usually conductive, that substantially reduces the effect of electric and/or magnetic fields.

Short circuit An undesirable path of very low resistance in a circuit between two points.

Short-circuit protection Any fuse, circuit breaker, or electronic hardware used to protect a circuit or device from severe overcurrent conditions or short circuits.

Signal The event or electrical quantity that conveys information from one point to another.

Significant digit A digit that contributes to the precision of a number. The number of significant digits is counted beginning with the digit contributing the most value, called the *most significant digit* (leftmost), and ending with the digit contributing the least value, called the *least significant digit* (rightmost).

Silicon-controlled rectifier (SCR) A semiconductor device that functions as an electronic switch.

Single-scan function A supervisory type instruction that causes the control program to be executed for one scan, including input/output update. This troubleshooting function allows step by step inspection of occurrences while the machine is stopped.

Sink mode output A mode of operation of solid-state devices in which the device controls the current from the load. For example, when the output is energized, it connects the load to the negative polarity of the supply.

Snubber A circuit generally used to suppress inductive loads; it consists of a resistor in series with a capacitor

(RC snubber) and/or a MOV placed across the alternating current load.

Software The term applied to programs that control the processing of data in a system, as contrasted to the physical equipment itself (hardware).

Solid state Term describing any circuit or component that uses semiconductors or semiconductor technology for operation.

Solid-state switch Any electronic device incorporating a transistor, a silicon-controlled rectifier, or a triode alternating current semiconductor switch to control the ON/OFF flow of electric power is often referred to as containing a solid-state contact or switch.

Source mode output A mode of operation of solid-state output devices in which the device controls the current to the load. For example, when the output is energized, it connects the load to the positive polarity of the supply.

State The logic 0 or 1 condition in programmable logic controller memory or at a circuit input or output.

Station Any programmable logic controller, computer, or data terminal connected to, and communicating by means of, a data highway.

Storage bit A bit in a data table word which can be set or reset, but is not associated with a physical input or output terminal point.

Subroutine A portion of a larger user program which may be accessed and executed any number of times during a single scan of the programmable logic controller.

SUBTRACT A programmable logic controller instruction that performs the mathematical subtraction of one number from another.

Suppression device A unit that attenuates the magnitude of electrical noise.

Surge A transient wave of current or power.

Synchronous transmission A type of serial transmission that maintains a constant time interval between successive events.

Synchronous shift register A shift register where only one change of state occurs per control pulse.

Syntax Rules governing the structure of a language.

System A set of one or more programmable logic controllers, input/output devices and modules, and computers, with associated software, peripherals, terminals, and communication networks, that together provide a means of performing information processing for controlling machines or processes.

T

Tasks A set of instructions, data, and control information capable of being executed by a central processing unit to accomplish a specific purpose.

Terminal address The alphanumeric address assigned to a particular input or output point. It is also related directly to a specific image table bit address.

Thermocouple A temperature-measuring device that utilizes two dissimilar metals for temperature measurement. As the junction of the two dissimilar metals is heated, a proportional voltage difference is generated which can be measured.

Thumbwheel switch A rotating switch used to input numeric information into a controller.

Timed contact A normally open and/or normally closed contact that is actuated at the end of a timer's time-delay period.

Timer In relay-panel hardware, an electromechanical device which can be wired and preset to control the operating interval of other devices. In a programmable logic controller, a timer is internal to the *processor,* which is to say it is controlled by a user-programmed instruction.

Toggle switch A panel-mounted switch with an extended lever; normally used for ON/OFF switching.

Transducer A device used to convert physical parameters such as temperature, pressure, and weight into electric signals.

Transformer An electric device that converts a circuit's electrical energy into a circuit or circuits with different voltages and current ratings.

Transistor A three-terminal active semiconductor device composed of silicon or germanium, which is capable of switching or amplifying an electric current.

Transistor-transistor logic (TTL) A semiconductor logic family in which the basic logic element is a multiple-emitter transistor. This family of devices is characterized by high speed and medium power dissipation.

Transitional contact A contact that, depending on how it is programmed, will be on for one program scan every 0 to 1 transition, or every 1 to 0 transition of the referenced coil.

Transmission line A system of one or more electric conductors used to transmit electric signals or power from one place to another.

Triode alternating current semiconductor switch (triac) A semiconductor device that functions as an electrically controlled switch for alternating current loads. A component of alternating current output circuits.

TRUE As related to programmable logic controller instructions, an ON, enabled, or 1 state.

Truth table A table listing that shows the state of a given output as a function of all possible input combinations.

U

UNLATCH instruction One-half of a programmable logic controller instruction pair which emulates the unlatching action of a latching relay. The UNLATCH instruction de-energizes a specified output point or internal coil until re-energized by a LATCH instruction. The output point or internal coil remains de-energized regardless of whether or not the UNLATCH instruction is energized.

Ultraviolet-erasable programmable read-only memory An erasable programmable read-only memory which can be cleared (set to 0) by exposure to intense ultraviolet light. After being cleared, it may be reprogrammed.

V

Variable A factor which can be altered, measured, or controlled.

Variable data Numerical information which can be changed during application operation. It includes timer and counter accumulated values, thumbwheel settings, and arithmetic results.

Volatile memory A memory structure that loses its information whenever power is removed. Volatile memories require a battery backup to ensure memory retention during power outages.

W

Watchdog timer A timer that monitors logic circuits controlling the processor. If the watchdog timer, which is reset every scan, ever times out, the processor is assumed faulty, and is disconnected from the process.

Word A grouping or a number of bits in a sequence that is treated as a unit.

Word length The total number of bits that comprise a word. Most programmable logic controllers use either 8 or 16 bits to form a word.

Write Refers to the process of loading information into memory. Can also refer to block transfer, i.e., a transfer of data from the processor data table to an intelligent input/output module.

Z

Zone control last (ZCL) state instructions A user-programmed fence for zone control last state zones.

Zone control last (ZCL) state zones Assigned program areas which may control the same outputs through separate rungs, at different times. Each such zone is bound and controlled by zone control last state instructions. If all zone control last state zones are disabled, the outputs will remain in their most recent controlled state.

INDEX